ELECTRICITY PRICING

Regulated, Deregulated and Smart Grid Systems

ELECTRICITY PRICING

Regulated, Deregulated and Smart Grid Systems

Sawan Sen
Academy of Technology, Hooghly, India

Samarjit Sengupta
University of Calcutta, Kolkata, India

Abhijit Chakrabarti
Jadavpur University, Kolkata, India

CRC Press
Taylor & Francis Group
Boca Raton London New York

CRC Press is an imprint of the
Taylor & Francis Group, an **informa** business

CRC Press
Taylor & Francis Group
6000 Broken Sound Parkway NW, Suite 300
Boca Raton, FL 33487-2742

First issued in paperback 2017

© 2015 by Taylor & Francis Group, LLC
CRC Press is an imprint of Taylor & Francis Group, an Informa business

No claim to original U.S. Government works

ISBN-13: 978-1-4822-5174-6 (hbk)
ISBN-13: 978-1-138-07401-9 (pbk)

Library of Congress Cataloging-in-Publication Data

Sen, Sawan.
   Electricity pricing : regulated, deregulated and smart grid systems / Sawan Sen, Samarjit Sengupta, Abhijit Chakrabarti.
      pages cm
   Includes bibliographical references and index.
   ISBN 978-1-4822-5174-6 (hardback)
   1. Electric utilities--Rates. 2. Electric utilities--Government policy. 3. Electric utilities--Deregulation. 4. Smart power grids. I. Sengupta, Samarjit. II. Chakrabarti, Abhijit. III. Title.

HD9685.A2S43 2014
333.793'231--dc23   2014015078

Visit the Taylor & Francis Web site at
http://www.taylorandfrancis.com

and the CRC Press Web site at
http://www.crcpress.com

Dedicated to our parents

Contents

List of Figures

List of Tables

Preface

Electricity is the prime mover of a modern society. Per capita consumption of electricity is an indicator of a country's growth. This indicator depends on availability, quality and reliability of power and most particularly its overall cost. The cost of electricity governs its usage. Thus, electricity pricing is a major issue to consumers for its liberal use. But due to demand of social and technological advancements, the changes in power networks and their mode of operation are inevitable. Revolutionizing electric power industries encompasses technical and also various non-technical and economic issues. Still, as an inevitable consequence, power networks have been undergoing changes from regulated to deregulated structures, finally stepping into the smart grid concept, which in turn involves a change in electricity price methodology. For maintaining reliable power available in deregulated and smart grid environments, recurring and non-recurring expenditures in different heads components have increased a lot compared to a regulated system, resulting in a high cost of electricity. To limit this cost within the reach of consumers is thus the chief motivation of research in this area, i.e. optimizing electricity pricing in modern power networks.

This book aims to portray the route of transformation of power networks from regulated to deregulated and finally the smart grid environment, the changes in electricity pricing due to this transformation, and in conclusion, different solution methodologies for optimizing electricity cost with respect to system performance and benefits of both power investors and consumers. As power network performance and electricity cost complement each other, the book starts with power system performance analysis for a regulated system, followed by some techniques for its improvement. This has revealed the limitations of the investment cost of the system and the necessity of deregulation. Optimum operation methodologies have been discussed with their effects on electricity pricing. In the second half of the book, the benefits of a smart grid have been conferred, followed by its possible economic operating conditions. In doing so, a number of innovative algorithms have been developed and presented that may be a guideline for practicing engineers to expand a modern power network that will produce cost-optimized quality power. A futuristic view of electricity pricing in the smart grid environment has also been drawn that may attract present and future engineers for further research.

This book is the outcome of research carried out by the authors at University of Calcutta, Department of Applied Physics, West Bengal, India. The contents have been prepared such that the book will serve the purposes of graduate, post-graduate and research students. A large number of references have also been given for further reading for those who are interested.

The authors are indebted to the faculty members of the Department of Applied Physics, University of Calcutta, and the Department of Electrical Engineering, Bengal Engineering and Science University, West Bengal, India, for their cordial suggestions during preparation of the manuscript. The authors also thank a number of practicing engineers of different utility companies for sharing their practical experiences. The authors wish to put on record the sincere cooperation offered by the associated team members of CRC Press.

And last but not least, the authors owe a debt of gratitude to their family members, whose continuous endurance helped this book to see the light.

Sawan Sen
Samarjit Sengupta
Abhijit Chakrabarti

About the Authors

Sawan Sen received her B.Sc. (Hons.) degree in physics and B.Tech. in electrical engineering from the University of Calcutta in 2001 and 2004, respectively. In 2006 and 2012 she obtained her M.Tech. degree and Ph.D. degree in electrical engineering from the University of Calcutta, India. She is currently working as associate professor of the Academy of Technology (Electrical Engineering Department), West Bengal. Her main research interests include power system stability analysis, system performance enhancement and different soft computing techniques for solving power system problems like congestion management, cost optimization and electricity pricing under regulated, deregulated and smart grid environments.

Samarjit Sengupta received his B.Sc. (Hons.) degree in physics and B.Tech, M.Tech and Ph.D. degrees in electrical engineering from the University of Calcutta, Kolkata, West Bengal, India. He is currently a professor of electrical engineering with the Department of Applied Physics, University of Calcutta. He has to his credit 130 journal papers and 8 books on various topics of electrical engineering. His main research interests include power quality instrumentation, power system stability, and security and power system protection. He is a Fellow of IET(UK), IETE and a senior member of IEEE(USA).

Abhijit Chakrabarti, born in West Bengal, India received his Ph.D. (Tech) degree from the University of Calcutta India in 1991. Currently, he is the vice chancellor of Jadavpur University, Jadavpur, Kolkata and a professor in the Department of Electrical Engineering, Bengal Engineering and Science University, Shibpur, Howrah, West Bengal, India. He has served on the West Bengal State Council of Higher Education as vice chairman and chairman (acting) during 2012–13. He has authored 123 research papers including publications in international journals published by Elsevier, IEEE

Transaction, AMSE, etc. Dr. Chakrabarti has authored 12 books on power systems, circuit theory, basic electricals, power electronics and electrical machine design. He has also guided a number of Ph.D. scholars. He received the Pandit Madam Mohan Malviya Award, Power Medal and an award for the best paper in research related to power system engineering. He is a Life Fellow of IEs (l) and he has executed several research projects from different government funding agencies in India. He is a member of different expert and policy making committees at various universities and the All India Council for Technical Education (AICTE), including the National Board of Accreditation (NBA) India. He has active interests in electrical power systems, electrical machines, and electrical circuits and power electronics.

List of Principal Symbols

C_{TOTAL}	Total generation cost
C_{0i}, C_i^j	Generation cost of each generator for the base and contingent state transaction
CC_i^j	Cost of unavoidable congestion
C_{TD}	Total cost of generation with the changed demand response (DR)
C_T	Total generation cost with flat DR
$\Delta\theta$	Incremental changes in bus voltage angle
δ	Voltage angle of bus
λ	Loading factor
ϕ	Activation function operator of ANN
α	TCR firing angle
η	Learning rate of training procedure of ANN
Σ	Diagonal matrix of positive real singular values σ_i
Σ^2	Eigen values of matrix JJ^T
$[\xi]$	Right eigenvector of $[J]$
$[\eta]$	Left eigenvector of $[J]$
$[J]$	Jacobian matrix
J_R	Reduced VQ Jacobian matrix
J_{max}	Maximum value diagonal elements of Jacobian matrix
J_{min}	Minimum value diagonal elements of Jacobian matrix
LCP^j	Active power load curtailment in MW
LCQ^j	Reactive power load curtailment in MVAR
LC_i^j	Cost of curtailed load
NG	Number of generators
NB	Number of buses
ND	Number of demands
P_{LLM_i}	Local load margin
P_L	Active loss
$\Delta P_{Gi}^j, \Delta Q_{Gi}^j$	Active and reactive power redispatched in MW and MVAR
$P_{G_i}^{min}, P_{G_i}^{max}$	Minimum and maximum active power limit of ith generator output

$\Delta P_{Gi}^{\min}, \Delta P_{Gi}^{\max}$	Minimum and maximum allowable active power redispatched in MW
P_D, Q_D	Total active and reactive power demand of system
P_L, Q_L	Total active and reactive power loss of system
P_G	Power generation of generator
P_{Gi}^0, P_{Gi}^j	Power generation of ith generator before and after contingency
$P_{ij\,\max}, P_{ij\,\min}$	Maximum and minimum active power flow between the buses i and j
P_{ij}^0, P_{ij}^j	Line flow between bus i and bus j before and after contingency
P_i^0, P_i^j	Total power injected to bus i before and after contingency
$P_l^{\max}$	Maximum allowable limit of active load curtailment
$\Delta P_{di}, \Delta Q_{di}$	Change in active and reactive power demand of ith load as informed by DR
ΔQ_{PQ}	Incremental changes in reactive power of PQ buses
$Q_{Gi}^{\max}, Q_{Gi}^{\min}$	Maximum and minimum reactive power limit of ith generator output
$\Delta Q_{Gi}^{\max}, \Delta Q_{Gi}^{\min}$	Maximum and minimum allowable reactive power redispatched in MVAR
Q_L	Reactive loss
$Q_{li}^{\max}, Q_{li}^{\min}$	Maximum and minimum allowable limit of reactive load curtailment
Q_{load_j}	Reactive load at bus j
S	Forecasted range of common bidding price of consumer and supplier
sf	Scaling factor
$[U]$	Diagonal eigenvalue matrix of $[J]$
V	Voltage magnitude of bus
$V_i^{\max}, V_i^{\min}$	Maximum and minimum voltage magnitude of bus i
ΔV	Incremental changes in bus voltage magnitude
W_1	Weight matrix between the input layer and hidden layer
W_2	Weight matrix between the hidden layer and output layer

List of Abbreviations

ABB	Asea Brown Boveri
AF	Acceleration factor
AI	Artificial intelligence
ANN	Artificial neural network
A/S	Ancillary services
CI	Curtailment index
CIGRE	Conseil International des Grands Research Electriques
CSF	Contingency sensitivity factor
DE	Differential evolution
DER	Distributed energy resource
DISCO	Distribution company
DP	Dynamic programming
DR	Demand response
EBNN	Elman backpropagation neural network
EHV	Extra-high voltage
ELD	Economic load dispatch
FACTS	Flexible AC transmission system
FC-TCR	Fixed capacitor thyristor-controlled reactor
FDLF	Fast decouple load flow
FVSI	Fast voltage stability index
GA	Genetic algorithm
HVAC	High-voltage alternating current
HVDC	High-voltage direct current
IEE	Institute of Electrical Engineers
IEEE	Institute of Electrical and Electronics Engineers
IPC	Interphase power controller
ISO	Independent system operator
LDC	Local dispatch centre
LMP	Locational marginal price
LPS	Longitudinal power supply
LQF	Line quality factor
LSE	Load serving entity

LTC	Load tap changer
MCP	Market clearing price
MOPSO	Multi-objective particle swarm optimization
NR	Newton Raphson
OI	Overloading index
OLF	Optimal loading factor
OPF	Optimal power flow
PCA	Principal component analysis
PEV	Plug-in electric vehicle
PML	Point of maximum loadability
POPF	Probabilistic optimal power flow
PS	Phase shifter
PSO	Particle swarm optimization
PTDF	Power transfer distribution factor
PWM	Pulse width modulation
RGC	Rate of generation cost
RTO	Regional transmission organization
SCADA	Supervisory control and data acquisition
SIL	Surge impedance loading
SO	System operator
SSI	Security sensitivity index
SSSC	Static synchronous series compensator
STATCOM	Static compensator
SVC	Static VAR compensator
TCR	Thyristor-controlled reactor
TCPS	Thyristor-controlled phase shifter
TCSC	Thyristor-controlled series compensator
TRANSCO	Transmission company
UDC	Utility distribution company
UPFC	Unified power flow controller
VCPI	Voltage collapse proximity indicator
VIS	Vertically integrated system
VSC	Voltage source converter
VSI	Voltage stability index
VOCC	Value of congestion cost
VOEL	Value of excess loss
VOLL	Value of lost load

1

Prologue

1.1 Motivation of the Book

The restructuring process in the electric power industry over the last decade has led to several structural and regulatory issues regarding transmission grid operation and planning, not fully anticipated at the design state of the grid. The transmission system has not evolved at the rate needed to sustain increasing demand matched with negligible generation addition evidenced in the deregulated environment. This has caused somewhat unexpected bottlenecks like voltage instability, low operating efficiency, poor loadability, price volatility, frequency droop and congestion in the system. Moreover, the functional unbundling of generation and transmission operations is aggravated due to the lack of coordination between generation resources and the transmission system operator (SO). In this situation, the role of SO is not only to increase the profit margin of the market participants but also to improve the performance of the network for consumer welfare.

The inextricably complex and interconnected structure of the grid is the main hindrance to improving its performance. Unavoidable circumstances like voltage instability and frequency droop are quite evident in the dynamic response of the system. Their gruesome effect can cause the collapse or cascading failure of the system. In this respect, line congestion is not also lagging behind. A single-line contingency can overburden the system to cause its failure even though all the generators are operational. Again, voltage instability according to the research conducted in this field has an intimate relation with angle stability, which further attracts frequency instability in the system. The loadability of a line is directly related to its voltage; hence, voltage instability determination and remedial action can shun severe effects of both frequency instability and poor loadability. Both in the long and the short run, poor efficiency can increase the marginal cost of the energy, which has dreadful effects on the economical aspect of power generation. The ever-increasing numbers of constraints are also complicating the objective of optimal power system operation toward the most economical pattern of parameters. Hence, voltage instability, poor efficiency, sub-optimal operations and congestion are the four basic dimensions of an ill-performing system.

In this context, it is increasingly important for the system operator to project and assess the operating condition of the system in terms of the above constraints and to adopt corrective actions to minimize their deviations to maintain recommended and desired performance of the system. For this, the SO would require the assistance of novel methodologies and algorithms to plan the most beneficial operating condition of the system, to model the network for optimizing the operating efficiency and for the maintenance of constraints within safe operating limits.

These reasons have motivated power engineers of the last decade to develop modeling methods, optimization algorithms and constraint management techniques for upgrading the role of the SO in the modern market in the transition of getting deregulated. However, the efficiency of these works has been influenced by aggravated inaccuracy, a one-dimensional objective, inefficient estimation capacity and time complexity. The search for more accurate, computationally simple, multi-objective fast response and robust algorithms still continues.

1.2 Contributions of the Book

In pursuit of multi-dimensional flexible and efficient methodologies, research works have focused on the steady-state and dynamic behaviour of the modern power market in an endeavour to optimize the system operating conditions. The main achievements of the book are as follows:

- The theoretical modifications of different voltage stability indices in practice have been adopted for implementation. For accurate and fast dynamic response, artificial neural network (ANN)-based training methodologies have been developed. The conventional stability assessment technique has thus been modified for fast and accurate monitoring of the system and for the development of real-time ancillary service for the generation of actuating signals to control the voltage stability of the system.

- Novel and effective modeling of compensation devices has been carried out in pursuit of maximizing the utilization of network capacity. ANN-based flexible AC transmission system (FACTS) models have also been developed for faster prediction of the control parameter for effective redemption. But, installing of different FACTS devices for improvement of system performance incurs more investment. In the case of the regulated power market, such types of investment may not be permitted, since they increase electricity prices. This argument compels the power engineers to move from a regulated power market to a deregulated power network where power investors are allowed to solicit higher electricity prices from consumers to maintain the quality of the power.

- Hence, a suitable contingency analysis technique has also been implemented to establish the effectiveness of the FACTS and high-voltage direct current (HVDC) links under regulated and deregulated environments for the restoration of the system operating point to a stable mode of operation. The techniques developed will assist the system operator in judging and placing compensating devices in optimal locations to increase the reliability and efficiency of the system.

- To utilize the stability margin and optimum network capacity, the effort has focused on novel methodologies to be undertaken by the system operator to optimize consumer welfare in terms of both electricity pricing and quality of the power. In this respect, the work has contributed in the field of cost optimization to minimize the price volatility of the power market. For the development of a global search algorithm for an optimal generation schedule, the work has deployed stochastic methods like genetic algorithm (GA), particle swarm optimization (PSO) and differential evolution (DE). Due to the uncontrollable rise of operating constraints, the objective function has become highly non-linear, and hence the amalgamation of these mathematical techniques can reduce the anomalies of market price in the modern deregulated power market.

- The modern-day power networks are outstressed by the irrepressible rise of demand. Line congestion is an evident effect of outstressing the network. The management method employed in this application has shown to be capable of sustaining the level of congestion. Non-derivative-based optimization techniques have also been deployed in this respect.

- The development of optimization techniques went further to enable the independent system operator (ISO) to redistribute generation for effective upgradation of transmission efficiency. In modern power markets, these algorithms can be effective for a high profit margin with consumer relief.

- Electricity, considered by most to be energy, is actually an energy currency. Power from a variety of sources, such as water, fuel, wind and solar, is used to create electricity for delivery to the consumer's demand. Thus, electricity has proven to be a suitable and competent means of delivering energy. But, there is no means to store electricity without converting it to another form of energy. As a result, the demand for power, driven by users, must match the supply of power from the available sources (e.g. thermal power and all other renewable energy sources) at all times. Thus, the modern-day power markets can be planned to be equipped with efficient smart meters, and work has employed these data collected by the meters to assess the present operating condition not only to reschedule generation but also to redistribute the demand for the overall improvement of system performance. The system operators of modern power networks can benefit from producing fast and accurate decisions for the maintenance of most suitable operational modes. This directs the corridor of application of the smart grid concept in the power network. 'The Smart Grid is ultimately about using megabytes of data to move megawatts of electricity more efficiently and affordably' (Ontario Smart Grid Forum report, May 2011). The study of smart grid has also found that by giving users a choice of when they use energy, and how much and what kind of energy they consume, they are getting more use out of green power sources and consumers are actively changing how they use the system, with changing electricity price.

1.3 Organization of the Book

This book includes seven chapters, including this prologue. Its organization is given below.

In Chapter 2, background and literature review are presented. This chapter starts with a general introduction to present the power market scenario and the power network performance evaluation techniques. The importance of voltage stability on power networks and a thorough literature survey in this field have been depicted. To illuminate the developed methods for stability assessment so far, classical and neo-classical techniques have been discussed. All these methods pointed toward a common solution – compensation – which indicates that more investment cost results in high electricity price; henceforth, the recent developments of FACTS and HVDC technology have been depicted. To emphasize the improvement of system performance, different optimization methods have been studied, which includes cost optimization strategies, congestion management techniques and loss optimization methods. The deployment of demand response (DR) for effective enhancement of system performance has also been highlighted with a meticulous survey of the works in this field.

The content of Chapter 3 is based on the analysis of voltage stability of longitudinal power supply (LPS) using an artificial neural network (for fast and accurate on-line analysis). The introduction of this chapter covers the developments of voltage stability assessment techniques. The theoretical progress of voltage stability and collapse indicators in the power sector has been discussed for their comparison and for the development of efficient indicators for proper assessment of the voltage stability margin as applicable to a practical system. This section of the book also deals with an elementary overview of ANN and the supervised training method adopted for the implementation of an ANN-based stability predictor. The results of this neo-classical technique have been compared with the classical technique for the assessment of accuracy and response.

In the quest for enriching system performance, Chapter 4 reinforces the utilization of FACTS devices and HVDC links. In the introduction, it depicts the present scenario of FACTS and HVDC deployment in the power sector. Elaborating the recent developments of the FACTS controllers, like static VAR compensator (SVC), thyristor-controlled series compensator (TCSC), etc., the effort concentrates on the modeling of these controllers for on-line prediction of control parameters to sustain the specified operating conditions. The HVDC system, being one of the modern invasions of power electronics into power systems, incurs high installation cost but offers high reliability. Henceforth, the work employed HVDC controllers to sustain the same operating conditions that can be done by FACTS controllers. These comparisons will enable the system operator to choose proper aid for the network so far as compensation is concerned.

Chapter 5 deals with the multi-objective optimization algorithms to upgrade the performance of modern deregulated power markets. The transition from LPS to deregulation with an overview of the deregulated market has primarily been introduced in the beginning of this chapter, highlighting the operational constraints. In the presence of these constraints, the objective function becomes highly non-linear and thus can only be solved with stochastic methodologies like GA, PSO and DE, which have been illustrated and compared for implementation in the present-day power market for maximization of utilities. For a multi-objective solution, the work focuses on the development of the same in different optimization environments. Power loss, generation cost and load curtailment optimizations for congestion management have also been discussed. To sustain the reliability of power supply, without breaching the limits of operational constraints, an optimization model has also been developed to manage congestion without load curtailment. For contingency surveillance, some indices have also been formulated that can assist the independent system operator (ISO) to control the limit violation incurring by the power system.

Apart from utility maximization, with the available smart metering system and supervisory control and data acquisition (SCADA), the system operator can schedule the loads of an interconnected network and some algorithms can be developed to improve the performance from the demand side. This is the topic of discussion of Chapter 6. Elaborate depictions of innovative methodologies have been presented to utilize demand response (DR) for maximization of profit as well as consumer welfare in the smart grid arena. This chapter also endeavours to portray an outline for estimating the cost and benefits of the smart grid.

Finally, Chapter 7 contains the conclusion and future scope of efforts described in the preceding chapters of this book.

2

Background and Literature Survey

Due to socio-economic reasons, the existing vertically integrated system is getting converted to deregulated systems. But the basic problems like catering extensively high demand and managing voltage stability and system efficiency remain the same along with economical considerations. This chapter deals with a thorough and meticulous survey of existing and projected technologies in the field of power engineering, to ensure safe and reliable operation of the system. The bright prospect of high-voltage direct current (HVDC), flexible AC transmission system (FACTS) and optimization techniques has been a part of the quest made in this chapter. The most recent developments like the smart grid have also been illuminated to extend this pursuit to future power networks.

2.1 Introduction

There are many factors involved in the successful operation of an electric power system. The system is expected to have power instantaneously and continuously available to meet consumer demands. Hence, in order to meet the demands, it is necessary to commit adequate generating units. The most economical operation of the modern power market demands proper interaction of the major control functions such as economic dispatch, performance analysis, etc. An overall solution to these sets of problems must result in a continuous and reliable supply of electricity while maintaining the optimal cost of production and desired operating conditions. Under the socio-economic changes where the regulated or state-owned monopoly market has been deregulated, it has become a challenging task to improve the performance in terms of cost optimization and congestion management without breaching the stability limits and other operational constraints.

Henceforth, the purpose of this chapter is to familiarize the reader with the implemented as well as proposed technologies available for the improvement of system performance. One of the primary requirements in this regard is to assess the present performance of the system with classical and neo-classical techniques, which are presented in the beginning of this chapter. Implementations of FACTS controllers and HVDC links are discussed to motivate the system operator to implement these devices for their long-term benefits. Present-day optimization techniques are discussed with an objective of approaching the most efficient algorithms. For the enrichment of this chapter, the operating conditions, performance, cost optimization methodologies and pricing technologies of regulated and deregulated power markets have been studied [1–4].

2.2 Power Network Performance Evaluation

The study of these markets has pointed toward a few common solutions for power network performance evaluation and improvement. Maintenance of a considerable voltage stability margin, upgradation of transfer capacity by compensation and optimizing the generation and load schedule are among important keys for the enrichment of network utilization. These techniques have been discussed in the following sub-sections.

2.2.1 Importance of Voltage Stability on Performance Evaluation

Power utilities are now forced to increase the utilization of existing transmission facilities to meet the growing demand and for the improvement of power network performance. Maintenance of voltage profile and stability is important to maximize the use of the network for catering to the request of a particular region at high efficiency [5]. Voltage instability is characterized by the inability of the system to retain its voltage near the nominal value, even with a change in connected susceptance at the load bus. To maintain a standard operational condition, the voltage stability should be vividly monitored by the system operator. Research on the voltage instability of an interconnected power system has been ongoing [6, 7]. The methods adopted for the determination of stability have turned up in different literatures and are furnished below.

2.2.1.1 Classical Methods of Ascertaining Stability

J. Deuse et al. show some examples of dynamic simulations of voltage phenomena using a new general purpose stability program (EUROSTAG) covering the classical fields of transient, mid-term and long-term stability, and also the quasi steady-state conditions of a power system [8, 9].

Morison et al. discussed voltage stability analysis of power systems using static and dynamic techniques. Using a small test system, they presented results of time domain simulations to clarify the phenomenon of voltage stability for better understanding and modeling requirements. The same system was then analyzed using a static approach in which modal analysis was performed using system conditions (or snapshots, which approximate different stages along the time trajectory). The results obtained using the static and dynamic methods were compared and shown to be consistent [10].

However, system voltage instability can well be treated as a dynamic phenomenon [11–14], and in a weak power system, an increase in dynamic load or line tripping may lead to voltage instability. The voltage collapse phenomenon is also to some extent dependent on load characteristics. Therefore, load modeling plays a key role in voltage stability assessment [15, 16]. Researchers have proposed a number of techniques to analyze the voltage collapse phenomenon [17].

Conceptual and theoretical backgrounds of voltage instability problems have also been established [18–23] covering both the static and the dynamic aspect of the problem [24–26].

T. Lie et al. developed two methods of determining weak transmission stability boundaries based on the strong controllability and observability properties of power systems. For the group of generators being identified, the method described in [18] uses a coherency measure evaluated for the set of all inertial load flow contingencies to determine groups of coherent generators.

The problem of voltage control and voltage stability in a longitudinal power supply (LPS) system has attracted much attention since the last decade [27–32], but its occurrence might not have been directly linked with angle instability [19, 28]. Voltage stability, being one of the prime requirements for proper control and assessment of security of LPS systems, hinges on the coordinated response of all voltage and reactive power statuses throughout the network, and it is very sensitive to changes in real and reactive power demands [14, 31, 33]. Studies of voltage collapse [34] clearly indicate that voltage instability becomes almost certain following large contingencies.

R. Yokoyama et al. presented a flexible approach to a coordinated control of voltage and reactive power in order to enhance voltage security of an electric power system [35].

T. Van Cutsen and C. D. Vournas [34] reviewed the general methodology of analyzing voltage stability in the mid-term and in the transient timescale. They point out how the stability of mid-term dynamics can be predicted using constant power loads in transient timescale modeling.

Voltage stability has long been categorized as a phenomenon that could be investigated using load flow methods [36, 37]. The Newton–Raphson (N-R) load flow technique and fast decoupled load flow (FDLF) technique are widely accepted in load flow algorithm, and they can even be well utilized in order to check the system performance, including system stability on an off-line basis. The solutions may diverge once the power system is stressed [38]. Earlier research has also indicated that when the voltage collapse region is approached, the load flow algorithm converges slowly or not at all, and it becomes very difficult to find a step size to be used for the next iteration [30].

P. W. Sauer and M. A. Pai [36] have established a relationship between a detailed power system dynamic model and a standard load flow model.

The linearized dynamic model was examined to show how the load flow Jacobian appears in the system dynamic state for evaluating steady-state stability. Two special cases were given for the situation when singularity of the load flow Jacobian implies singularity of the system dynamic state.

The development of the physical concepts and mathematical backgrounds of voltage stability has been done on a basis of load flow solution feasibility [39], optimal power flow [40], bifurcation technique [41, 42], singularity of Jacobian [36], etc.

V. Ajjarapu and B. Lee [41] presented a tutorial introduction to bifurcation theory, and the applicability of this theory to study non-linear dynamical phenomena in a power network was explored.

C. A. Canizares [42] discussed the relation between bifurcations and power systems stability through a thorough analysis of several examples, to clarify some ideas regarding the usefulness and limitations of bifurcation theory in network studies and operation, particularly in voltage stability-related issues. The Ecuadorian National Interconnected System (SNI in Spanish) was used to depict and discuss the effect of load modeling in saddle-node bifurcation analysis of real power systems.

At the present time, it has been an accepted proposition that the singularity of the Jacobian in the load flow solution indicates a critical state of voltage [36], and the voltage stability index can be obtained from the feasibility of the solution to power flow equations for each of the buses [25] on off-line basis. The achievements lie in the domain of the static model as well as in dynamic voltage collapse models [28]. These models predict the proximity of the system near critical state and determine the reactive reserves, etc. However, the margin or proximity to the stability limit that is of concern to any system operator is not usually frequently explored. In the operation of a longitudinal power system (LPS), it is very pertinent to investigate voltage stability and security margins [17, 21, 30, 38, 43, 44]. Many researchers have proved that sensitivity analysis is an efficient tool to assess voltage stability [39, 45, 46]. Efforts have been made to determine suitable voltage stability indices to identify the weak/ weakest bus in the power system responsible for voltage collapse [18, 33].

T. Van Cutsen dealt with the diagnosis of voltage collapse situations, following large disturbances or load increases [46]. A method had been proposed to identify the set of buses where load restoration is responsible for the collapse and to determine the corresponding corrective actions. It was implemented in a fast voltage stability simulator, using sensitivity techniques. Tap changer blocking and load shedding were illustrated on a practical 410 bus system.

I. Musirin and T. K. Abdul Rahman [47] demonstrated the use of the line stability index termed the fast voltage stability index (FVSI) in order to determine the maximum loadability in a power system. The bus that is ranked highest in this method was identified as the weakest bus since it can withstand a small amount of load before causing voltage collapse. The point at which FVSI is close to unity was taken as an indicator of the maximum possible connected load and has been termed maximum loadability at the point of bifurcation. This technique was tested on the Institute of Electrical and Electronics Engineers (IEEE) test system and results proved the applicability of the developed technique to estimate the maximum loadability in a system.

PV and QV curves are commonly used to determine the steady-state voltage stability limit of a power system. M. H. Haque presented a new method [48] of determining the voltage stability limit using the PQ curve. For a given operating point, the voltage stability margin can easily be determined from the stability boundary in the PQ plane. The developed method of determining the voltage stability limit was tested on a simple system and very interesting results were found. PV or QV curves are also commonly used to determine the maximum permissible load (or static voltage stability limit) of a power system.

It has been observed from a literature survey that most of the authors used the Jacobian of the load flow equations as the workhorse for calculation of voltage stability; Kundur et al. [49] probably had given more weightages in using a reduced Jacobian matrix as well as a modal form of analysis for assessing voltage stability. Y.-H. Hong et al. presented similar work where they had used the minimum singular value of the Jacobian matrix of the load flow equation as an indicator of voltage collapse [50]. A. K. Sinha also depicted similar Jacobian-based proximity of the voltage stability indicator.

Other than employing the Jacobian matrix, M. Moghavvemi presented a line stability index-based voltage collapse indicator in [51]. Another collapse proximity indicator has been presented in [52]. Authors in [53–56] depicted steady-state stability-based indicators.

Optimal power flow (OPF) techniques have also been used to assess voltage stability [57]. Some researchers have incorporated contingency constraint in optimal power flow for proper voltage control in a power system.

2.2.1.2 Neo-Classical Methods of Ascertaining Stability

Due to the time complexity of classical approaches of determining stability, some cost-effective stochastic techniques were harnessed in the field of power engineering for faster on-line processing to ascertain stability. In the recent past a large number of research works have been developed for the solution of power engineering problems using ANN [58–62]. C. Dingguo and R. R. Mohler presented the possible consequences of not considering load dynamics, which at worst can be a complete voltage collapse. Based on this observation, modeling of load dynamics was considered in their paper, and neural networks (NNs), including recurrent neural networks, were applied for load modeling. Furthermore, they presented the strategies, for the first time, to incorporate the neural network-based load model into static and dynamic voltage stability analysis. The computation of the relevant sensitivity was carried out for the neural network-based load model, and the results were used in the popular modal analysis. The proposed methods were tested on both the IEEE 14 bus system and a practical system [63].

A multi-layer feed-forward artificial neural network (ANN) with error backpropagation learning has been proposed for the calculation of voltage stability margins by A. A. El-Keib and X. Ma [64]. The method efficiently determined the variation of the voltage stability margin in varying the demand state of the network. D. Paul et al. [65] presented an ANN function approximation-based training schedule of instability indicators, which utilizes the generator load mismatch and active/reactive power margins for on-line assessment of proximity of the voltage instability of a power network. Another feed-forward ANN-based novel approach of finding voltage instability by training with the L-index has been cited by S. Kamalasadan et al. in [66]. A combination of artificial intelligence (AI) technologies and three-dimensional PQV has been utilized by K. Yabe et al. in [67] for locating the point of voltage collapse of a large power network. H. B. Wan [68] presented a neural network-based approach for contingency ranking of voltage collapse by the singular decomposition method. He used a radial basis function (RBF) map to predict accurately, after training, the stability of a power network by its operating conditions. The effectiveness of ANN-based technologies has been extended to even deregulated power networks by B. Suthar when he cited a novel stability assessment method by neural network training in a combined pool and bilateral transaction mode regime in [69].

An enhanced RBF network-based multi-contingency voltage stability monitoring system was developed by S. Chakrabarti and B. Jayasurya [70]. The method effectively determines the power margin of different nodes in a large network for stability.

2.2.2 Significance of Compensation Techniques

The continuing interconnection of a bulk power system brought about by economic and environmental pressure has led to an increasingly complex system that must operate even closer to the limits of instability. Increased use of transmission facilities due to higher industrial output and deregulation of the power supply industry have provided the momentum for exploring new ways of maximizing power transfers in existing transmission facilities while at the same time maintaining acceptable levels of network reliability and stability. In this environment, performance control of the power network is mandatory. An in-depth analysis of the options available for achieving such objectives has pointed in the direction of power electronics [71]. Series capacitors are widely used in ultra-high-voltage networks in order to compensate for the series reactance of long lines [72–77]. Recently, they started to be used more in the distribution networks in Japan. Moreover, they are proposed to be used in arc furnaces feeding networks to increase production [75–77]. Like series capacitors, a shunt capacitor can also be used to artificially reduce the transmission distance and achieve maximum power transfer. In [78], shunt compensator modeling was illuminated for improving the voltage profile of a system. Amalgamation of these compensation techniques with power electronics has produced a new series of ancillary services of the power network known as FACTS.

There are two generations for realization of power electronics-based FACTS controllers: the first generation employs conventional thyristor-switched capacitors and reactors, and quadrature tap-changing transformers, and the second generation employs gate turn-off (GTO) thyristor-switched converters as voltage source converters (VSCs).

The first generation has resulted in the static VAR compensator (SVC), the thyristor-controlled series capacitor (TCSC), and the thyristor-controlled phase shifter (TCPS). The second generation has produced the static synchronous compensator (STATCOM), the static synchronous series compensator (SSSC), the unified power flow controller (UPFC), and the interline power flow controller (IPFC). Though these two groups of FACTS controllers have distinctly different operating and performance characteristics, generally they are able to change the network parameters in a fast and effective way in order to achieve better system performance [79–82].

The thyristor-controlled group employs capacitor and reactor banks with fast solid-state switches in traditional shunt or series circuit arrangements. The thyristor switches control the on and off periods of the fixed capacitor and reactor banks and thereby realize a variable reactive impedance. Except for losses, they cannot exchange real power with the system. The VSC type FACTS controller group employs self-commutated DC to AC converters, using a GTO thyristor, which can internally generate capacitive and inductive VAR for transmission line compensation, without the use of fixed capacitor or reactor banks. The converter with an energy storage device can also exchange real power with the system, in addition to the independently controllable reactive power. The VSC can be used uniformly to control transmission line voltage, impedance, and angle by providing reactive shunt compensation, series compensation, and phase shifting, or to control directly the active and reactive power flow in the line [83].

Although the FACTS controllers have copious advantages as compensating devices in power networks, HVDC is one more approach to improve the system performances. During contingency or fault HVDC can insulate two interconnected systems from each other, thereby improving reliability. Though HVDC has numerous technical advantages over FACTS, its use is limited by the high investment cost associated with it [84, 85], which is the most responsible factor for increasing electricity prices.

The following part of this chapter concentrates on the developments made in compensation and HVDC technologies.

2.2.2.1 Series and Shunt Compensation Employing FACTS Devices

For the purpose of this review, a literature survey has been carried out including two of the most important and common databases, the IEEE/ IEE electronic library and Science Direct electronic databases. The survey spans the last 15 years from 1990 to 2004. For convenience, this period has been divided into three sub-periods: 1990–1994, 1995–1999 and 2000–2004. The number of publications discussing FACTS applications to different power system studies has been recorded. The results of the survey are shown in Figure 2.1. It is clear that the applications of FACTS to different power system studies have been drastically increased in the last 5 years. This observation is more pronounced with the second-generation devices as the interest is almost tripled. This shows more interest for the VSC-based FACTS applications. The results also show a decreasing interest in TCPS, while the interest in SVC and TCSC slightly increases.

Generally, both generations of FACTS have been applied to different areas in power system studies, including optimal power flow [86–90], economic power dispatch [91], voltage stability [92, 93], power system security [94] and power quality [95, 96]. Applications of FACTS to power system stability in particular have been carried out using the same databases. The results of this survey are shown in Figure 2.2. It was found that the ratio of FACTS applications to the stability study with respect to other power system studies is more than 60% in general. This reflects clearly the increasing interest in the different FACTS controllers as potential solutions for the power system stability enhancement problem. It is also clear that the interest in the second generation of FACTS has drastically increased, while the interest in the first generation has decreased. The potential of FACTS controllers to enhance power system stability has been discussed by M. Noroozian and G. Anderson [97], where a comprehensive analysis of damping of power system electromechanical oscillations using FACTS was presented. H. F. Wang and F. J. Swift [98] have discussed the damping torque contributed by FACTS devices, where several important points have been analyzed and confirmed through simulations.

For the enhancement of power system performance, as discussed earlier, deployment of compensating devices has become quite imperative in modern power system networks. In [99], M. Z. El-Sadek et al. presented a steady-state

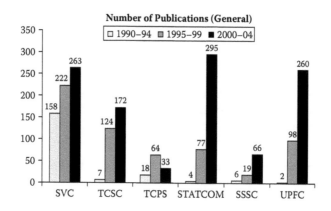

FIGURE 2.1
Statistics for FACTS applications to different power system studies.

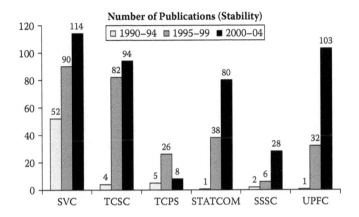

FIGURE 2.2
Statistics for FACTS applications to power system stability.

voltage stability enhancement technique with a combined effort of series capacitor and SVC. M. H. Haque in [100] depicted the fact that the voltage stability margin can be widened by employing SVC. For the improvement of dynamic performance, C. W. Taylor proposed a SVC model in [101]. The model proposed proved to be better than the model offered by CIGRE in [102]. One year later, however, the IEEE working group produced another model described in [103]. One of the crucial issues of deploying SVC is its placement. Reference [104] enlightens the critical understanding of optimal placement of SVC. The Newton–Raphson-based algorithm for reliable solution of large power networks with FACTS devices was proposed by C. R. Fuerte-Esquivel and E. Acha in [105]. The work efficiently models the FACTS devices with tap changer, phase shifter and series compensators. The same author extended his work in [106, 107] to optimize the models of TCSC and SVC, respectively, to meet the requirements of practical power networks. Several other examples of FACTS device modeling have been found in the literature [78, 108, 109]. In [110], faster prediction of voltage instability and restoration of the system by SVC was proposed by S. Dey. The work developed a unique global voltage security indicator for the assessment of voltage stability. Being associated with static switching devices, the response of these FACTS devices may be oscillatory, and to damp out this switching oscillation, a robust control strategy has been proposed by M. Noroozian et al. in [111]. C. R. Fuerte-Esquivel et al. implemented their model successfully to develop object-oriented power system software for the analysis of large-scale networks in [112].

2.2.2.2 Employment of HVDC Link

The profound utilization of the power electronic technology has reached the threshold of optimizing system performance by the incorporation of HVDC in power networks. The numerous aspects of improvement of system operating conditions have even left behind the high investment cost associated with it. In the present power market scenario, where different multi-national concerns are investing heavily for reliability and quality of power, HVDC can be a great choice. To increase distributed generation, specially in offshore wind power plants, HVDC can be taken as a better alternative of conventional transmission systems. Though on paper HVDC is profitable only for long-distance lines, different power systems around the globe are adopting this technology, even for back-to-back connection of grids. Power systems under development are strongly recommended for the procurement of HVDC links to encourage disperse generation and to improve reliability and quality of power to be transferred. It is a fact that if the future power system is going to work at its optimum stress, increase the sustainability of power, encourage clean energy and increase system efficiency, the HVDC link is the only solution. This has caused power system researchers to continuously project different ways of improving the system efficiency and reliability by the bulk use of HVDC.

In [113], the researchers have focused on the reliability statistics of HVDC technology. D. A. Waterworth in [114] went one step further to produce a technique to determine the reliability of these links. In [115], D. Jovcic et al. presented a new controlled strategy of the inverter side of HVDC to strenghten a weak HVAC system. The method cited comprehensively reduced the switching transient of the inverter side to upgrade the transient stability margin of the system. The researchers in [116] developed a non-linear control strategy for further improvement in the stability margin. In the work, G. J. Li et al. proposed and proved a pulse triggering technique that effectively reduces the transient time to aid stability.

Though this technique enhances the stability margin quite effectively, one of the primary hindrances of implementing HVDC in practical systems is its high investment cost. ABB Corporation took a great initiative and cited a few cost comparisons in [117] of HVDC systems for better marketing of the product to maximize transmission network facilities (Table 2.1). The work also depicted the recent development made to project the advantages offered by different HVDC topologies.

Siemens in [118] marketed its HVDC product and depicted the growing population of integration of HVDC links in the system worldwide (Figure 2.3).

H. F. Latorre and M. Ghandhari in [119] turned up with a design of a VSC-controlled HVDC link for the improvement of system stability. In his work, he has developed a technique to determine the most suitable location for the incorporation of an HVDC link.

An investigation of the improvement of voltage stability employing an HVDC link is presented in [120]. The authors in this paper have discussed the perturbation of PV and QV curves under the influence of a hybrid HVDC-AC link. Practical implementation of an HVDC-AC link in Australian and European grids was presented in [121, 122]. They simulated a practical system with an HVDC link to find numerous power flow solutions and wider stability margins.

TABLE 2.1

Cost Comparison of AC Transmission and Different HVDC Topologies

	AC Alternatives			Hybrid AC/DC Alternatives		
Alternative	500 kV Single	500 kV Double	765 kV Two Single	500 kV Bipole	500 kV Single	Total AC + DC
Rated power (MW)	3000	3000	3000	3000	1500	4500
Station costs including reactive compensation	542	542	630	420	302	722
Transmission line costs (M$/mile)	2.00	3.20	2.80	1.60	2.00	—
Distance in miles	1500	750	1500	750	750	1500
Transmission line cost (M$)	3000	2400	4200	1200	1500	2700
Total cost (M$)	**3542**	**2942**	**4830**	**1620**	**1820**	**3422**
Annual payment, 30 years @ 10%	376	312	512	172	191	363
Cost per KW-year	125.24	104.03	170.77	57.28	127.40	80.66
Cost per MWh @ 85% utilization factor	16.82	13.97	22.93	7.69	17.11	10.83
Losses at full load in %	6.93%	6.93%	4.62%	5.29%	4.79%	5.12%
Capitalized cost of losses @ 1500 KW(M$)	265	265	177	135	61	196

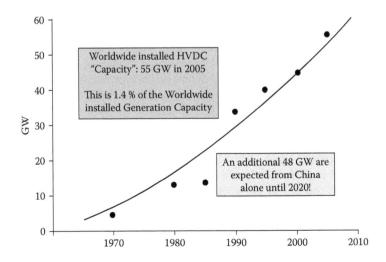

FIGURE 2.3
Worldwide installed capacity of HVDC link.

In recent works, the firing angles of HVDC converters have been determined by employing a stochastic algorithm like particle swarm optimization (PSO) for faster implementation and better transient response of the system [123].

2.2.3 Optimization Methods with System Performance and Cost Emphasis

In power systems, to maximize the cost-effective system performance under operational constraints a proper objective function has to be formulated and efficient analytical tools are to be designed to obtain global optima of these objectives. Hence, formulation of an objective function and choice of optimization technique are of equal importance. The endeavour of the researchers over the years has been to formulate the most desirable objective function and to search the solution with the most efficient optimization method. The incapability of classical methods (Lagrange's and Bellmon's) to obtain the optimal solution in a rough working plane has given birth to neo-classical techniques like swarm intelligence, fuzzy logic, genetic algorithm and evolutionary programming. The following sections enlighten the evolution of optimization algorithms to the improvement of power system network performance.

2.2.3.1 *Classical and Neo-Classical Optimization Methods*

The optimal system operations in general involve the consideration of economy of operation, system security, emission at certain fossil fuel plants, optimal releases of water at hydrogeneration plants, etc. All these considerations may make for conflicting requirements, and usually a compromise has to be made for optimal system operation. In the early 1970s dynamic programming (DP) was quite popular to obtain the solution of power system optimization problems. The expansion of an electric power transmission system was proposed by J. C. Kaltenbach et al., who employed DP to maximize the transmission extension in [124]. State-space modeling of power network constraints and a solution for the power network problem using an optimal control strategy have been successfully implemented by Y. N. Yu et al. in [125]. In [126], A. M. Sasson and H. M. Merrill utilized Lagrange's multiplier technique to optimize the solution of power system network problems. In [127], J. Nanda et al. used Fleture's QP method in the solution of the optimal power dispatch problem. More derivative-based optimization techniques have been reported in [128–133]. These classical techniques of obtaining a numerical solution of the power network problem have been summarized in [134]. All these techniques suffer from premature adoption of solutions; i.e. the obtained solution cannot get near to the global best possible solution in the working plane. This is basically due to their heavy reliance on the first derivatives of the constraints. To assess the accuracy of these optimization methods, in [135, 136] R. Billinton proposed some methodologies for the evaluation of reliability of the solution obtained by the classical optimization method.

In [137], J. Kennedy and R. Eberhart came up with a neo-classical technique of obtaining a solution to optimization problems inspired by social behaviour of birds and fishes. The technique, referred to as particle swarm optimization, soon became popular with power system researchers for its ability to reach a global solution with a minimum number of iterations and found its applications as a stochastic technique of optimization in [138–144]. The confinement of the solution within the generated population is the social metaphor of PSO, which does not let the algorithm produce the best optimal solution in the working plane.

The genetic algorithm (GA) endowed with mutation and crossover phenomena inspired from biological cell division can produce relatively new solutions to survive and be the fittest one to remain optimal. This upgradation of solution in every iteration attracted power system researchers to implement it for the best possible solution of the network problem. L. L. Lai in [145] proposed an improved genetic algorithm-based optimal power flow solution for optimizing the network problem in both normal and contingent states. The authors of [146–148] followed the same steps to obtain the most logical solution of network problems. The upgradation work of GA for power network problems was taken another step ahead when W. Yan introduced the hybrid GA interior point method to optimize reactive power flow. Though the reliability level of classical techniques has made them extinct, some numerical methods with classical operators are producing better results than even GA. In [149, 150] researchers used predictor corrector and conic quadratic formulation to establish the applicability of a modified classical operator to produce better results. Nevertheless GA can be more effective with non-linear problems and can handle a large number of constraints. Hence, its application is quite evident in [151–154].

To upgrade GA, R. Storn and K. Price introduced a differential operator to initialize population (probable solution) with logic (which was random in the case of GA) in [155]. The later works in this field followed differential evolution (DE) to search for the best possible solution in [156–160]. S. Rannamayan et al. [161] introduced opposition-based DE for the upgradation of the same. In [162], M. Metwally et al. modified the differential operator to produce better results. G. V. John in the same year [163] had introduced the quantum-inspired evolutionary algorithm.

One of the remarkable adaptations in the power system optimization problem is the inclusion of the century-old fuzzy logic. An elementary example of it can be found in the recent works of S. K. Bath, V. C. Romesh and D. Hur in [164–169].

2.2.3.2 Application of Optimization Methods in Regulated and Deregulated Power Networks

In recent years, the electricity industry has been undergoing unparalleled restructuring all over the world. The regulated or state-owned monopoly markets have been deregulated [170]. This process is intended to open the power

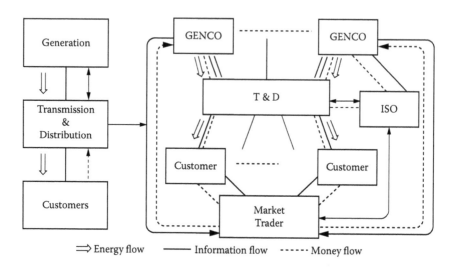

FIGURE 2.4
Conversion from VIS to deregulated power network.

sector to market forces with the ultimate target of reducing consumer prices. Therefore, the central ideology of electric power industry deregulation is that the delivery of power must be decoupled from the purchase of the power itself and be priced and contracted separately [171]. In these markets, the reliability and quality of power are priced and market participants rely heavily on the system operator to decide these prices. To ensure high reliability, the same transmission network is being used by different market participants simultaneously and is predicted to operate at its optimum design limit. Hence, line congestion [172] has become almost inevitable where lines of insufficient power flow capacity have to accommodate all the requests of the market participant. This line congestion and utilization of an already congested line for a new request will charge the participant with congestion management cost to avoid congestion by rescheduling. The conversion of a vertically integrated system to a deregulated system, shown in Figure 2.4, will increase the reliability, but at the cost of price volatility, line congestion and excess power loss.

The next section highlights the developments in the field of cost optimization – congestion management and power loss minimization strategies intended to improve the overall performance of the network in deregulated scenarios of power systems. After the inclusion of deregulation, the objective functions become more complex and non-linear; hence, the optimal power flow solution relies more on optimization techniques like PSO, GA and DE.

2.2.3.2.1 Cost Optimization Strategies

The main aim of the economic load dispatch (ELD) problem is to minimize the total cost of generating active power at various stations while satisfying the loads and losses in transmission links. If the specified variables are allowed to vary in a region constrained by practical considerations, there results an infinite number of load flow solutions, each pertaining to one set of specified values; the best choice in some sense of the values of specified variables leads to the best load flow solution. Economy of operation is naturally predominant in determining the best choice, though there are other factors that should be given consideration.

These issues have been highlighted by the IEEE committee in [173]. In [174, 175], T. W. Berrie highlighted power system economics and planning considerations. Optimal power flow solutions to the load dispatch problem have been illuminated in [176–178]. Reference [179] cited optimal maintenance scheduling of generating units. In a multi-objective optimal thermal power dispatch schedule, J. S. Dhillon et al. [180] included transmission line flow constraints in the economic load dispatch problem.

The challenges to on-line OPF implementation have been presented by J. A. Momoh et al. in [181]. In the work, several control strategies for real-time implementation of optimal power flow solutions have been cited. This work was developed in [182], where R. Bacher, the pioneer of this field, depicted automatic generation control for real-time optimum power flow. Implementation of stochastic methods for optimal power flow has been found in [183–185]. Some more instances of power flow solutions can be found in [186–190].

One of the constraints of real-time pricing of electricity is that it requires efficient monitoring and control system, which is of course costly and requires faster determination of price. In [191], Goldberg presented a genetic algorithm-based search optimization. Optimal spot pricing of electricity has been proposed by F. C. Schweppe et al. [192]. The method implemented effectively reduced the constraints of the deregulated systems to simplify the cost optimization process. In [193], the same author illustrated different spot pricing practices round the globe to compare their effectiveness in social welfare. Another approach to generating price signals is the determination of locational marginal price (LMP). Different LMP calculations have been presented in [194]. A short-run marginal price-based active and reactive power production has been proposed in [195]. The method proposes a comparatively faster calculating program to determine the price signals. Similar approaches of determining optimal spot price can be found in [196–200]. A power transfer distribution factor (PTDF)-based optimization model has been illustrated in [201], where PTDF is employed for allocating power transactions in deregulated markets. A hybrid model of electric power spot price presented by Matt Davison illuminated the basic difficulties of controlling price spikes in deregulated markets. Related works carried out earlier in this field can be found in [202–209]. All these optimal power flow models aiming for maximization of utilization of available resources point toward a common solution: compensation. A compensation technique for optimal choice and allocation of FACTS devices has been presented in [210]. Different FACTS-based optimal power flow models have been proposed in [211–214]. The research work in the field of cost optimization is still going on with the aim of maximizing the capability of power networks. But day by day the constraints are increasing (pollution, emission, congestion) and the objective function is getting more and more complex. Hence, an absolute formulation of objective function for OPF is imperative and still to be designed in the near future.

2.2.3.2.2 Congestion Management Strategies

Transmission network congestion is the main constraint to optimum exploitation of energy sources. Congestion can be caused by transmission line and generator outages, changes in energy demand and uncoordinated transactions, which can lead to network congestion when the transmission system is not able to respect security requirements due to line overload, transient and stability constraints [215, 216]. Looking for increased competition on electric power markets, the industry restructuring process tends to deepen the congestion problem [217]. Therefore, some schemes of congestion management are to be implemented that will internalize the dispatch process with externalities without increasing the electricity prices unreasonably and will keep the motivation for the market players to make investments on system expansion. A modal participation factor-based congestion management algorithm was developed in [218] for effective reduction in line congestion by monitoring the Jacobian matrix and eigenvalues of the system. Authors of [219] came up with a hybrid model with the curtailment strategy to optimize generation cost in a deregulated fuzzy environment with congestion management.

Line congestion apart from causing limit violation can cause a serious rise in spot price. Y. Peng et al. [220] has depicted different control mechanisms to limit spot price by congestion management. He presented some congestion management methods, namely generation control, demand control and FACTS control. Mitigation of congestion by employing FACTS devices can be found in some recent works [223–227]. Reliability evaluation of hybrid power markets during congestion has been depicted in [228]. It discussed an optimal load curtailment strategy managing congestion. Optimum load shedding-based congestion management was also found [229], but due to the loss of reliability associated with it and the installation cost involved with FACTS devices, researchers came up with alternative strategies like bidding control, generator and load contribution factor-based management, generation rescheduling and dynamic congestion management. An extensive endeavour in the deployment of this technique can be found in [230–238]. As the modern electricity market is getting deregulated, congestion management should cost the market players, so that they can be more conscious about avoiding transactions causing line limit violations. H. Yang and M. Lai presented a bilateral model for congestion cost evaluation in [239].

As different transmission companies have embarked upon providing a for-profit transmission service to market players for the changing power industry, estimation and projection of the transmission congestion range have become crucial. The task is complicated due to the difference in the congestion management and pricing protocol adopted by different electricity markets as well as the lack of relevant data posted by ISO. The congestion rent calculation for the California and New York electricity markets is summarized in [240]. The lack of public information concerning line power flows in the grid has made it difficult to estimate the total transmission congestion cost to be collected by ISO. Accordingly, sophisticated tools such as a probabilistic optimal power flow (POPF) are needed to accomplish the task. Unfortunately, numerous uncertainties associated with the behaviour of the market players and the evolution of the power system topology, coupled with the computational efforts, severely limit the successful application of POPF for this purpose. Hence, development of the POPF algorithm for congestion management has become a great challenge for the power engineers of the 21st century.

2.2.3.2.3 *Improvement of Transmission Efficiency*

Transmission loss incurs roughly 3–5% of the total generation to be considered as one of the major factors in locational spot pricing. This means that loss allocation may considerably affect the competitive position of the GENCOs. Nevertheless, it seems that most of the electric markets hardly ever reflect the transmission loss in their spot pricing due to its complicated aspects of loss allocation, such as non-linearity, path dependency and non-uniqueness of the solution. The important issues are to reduce allocation error, that is the discrepancy between the sum of theoretically allocated losses and the actual system loss. Efficient loss allocation and loss optimization methods can upgrade the transmission efficiency not only to improve performance but also for consumer welfare by lowering the price volatility of the electricity market. Several models of loss optimization have been proposed and are still being proposed by the power engineers to improve the loss profile of a system. An OPF tool with an objective of minimizing transmission loss can found in [241]. The approach has been followed by numerous developments [242–244]. But, the idea of loss optimization can prematurely converge to a sub-optimal solution for the constraints involved in modern power networks. The transmission loss allocation approach has upgraded to the power loss optimization model. S. Ahamed proposed a loss allocation model in the deregulated electricity market to minimize the physical flow in transmission lines by reducing the contribution of individual loads to the losses [245]. The work is inspired by the developments of T. K. Hann et al. for calculating the transmission loss factor found in [246]. The loss allocation approach is still under development, and hence different strategies can be found in [247–250]. The contributing factors of transmission loss are not only the active power requirement of the market players, but also the reactive power requirement. Optimal reactive power flow and consistence management of the same are highly desirable with systems conceding more cost to produced power to be lost. A reactive power management proposal can be found in [251]. In [252] a stochastic approach with GA to optimize reactive power under deregulation was used. Reactive power allocation as found by these researchers is a very complicated task, and the direct solution cannot be obtained with conventional algorithms and methodologies.

The last decade has witnessed drastic changes in the electric power industry in many parts of the world. The vertically integrated system (VIS) has been restructured and unbundled to one or more generation companies, transmission companies and a number of distribution companies. The implementation of deregulation in power systems is based on two main concepts: power pool and bilateral contacts. In both cases, a transparent method for allocating transmission losses between all of the interested parties in an equitable and fair manner is required. Hence, some methodologies and algorithms are required for loss optimization, and if not possible, then proper allocation of the same.

2.2.4 Enrichment of Cost-Governed System Performance in Smart Grid Arena

The huge interconnected machines spanning a large area of the globe are the century-old power grids, which are massively complex and inextricably linked to social and economic activities [253]. These grids are designed to connect large power stations to high-voltage transmission systems, which in turn supply power to medium- and low-voltage local distribution systems [254]. Due to socio-economic reasons, the growth of electricity demands requires reliability and quality of power supply, and for the maintenance of the same, the power industry is now facing unprecedented challenges and opportunities.

The smart grid technologies are currently undergoing development in an effort to modernize legacy power grids to cope with increasing energy demands without violating the above constraints [255–257]. It has been shown by researchers that the transmission, distribution and end users can be interconnected by high-speed bi-directional communication networks to optimize the grid operation. Integration of automation, supervisory control and data acquisition (SCADA) and smart metering in power networks can enable the grid to rapidly self-regulate and heal to improve system reliability and security [258].

The vision and different logical domain of a smart grid is illustrated in Figure 2.5. As a roadmap for research and development, the smart features of the transmission grid are envisaged and summarized in the figure as digitalization, flexibility, intelligence, resilience, sustainability and customization. With these smart features, the future transmission grid is expected to deal with the challenges in all identified areas.

For the efficient operation of smart grids, researchers have sought more participation of the consumers in scheduling of power of different sources. So far there has been significant research integrating consumer demand side management into the smart grid to improve the system load profile and reduce peak demand [259]. Demand response (DR) allows consumer load reduction in response to emergency and high price conditions on the electricity grids. Such conditions are more prevalent during peak loads or congested operation. Non-emergency demand response (DR) can provide substantial benefits in reducing the need for additional resources and lowering the real-time electricity price or

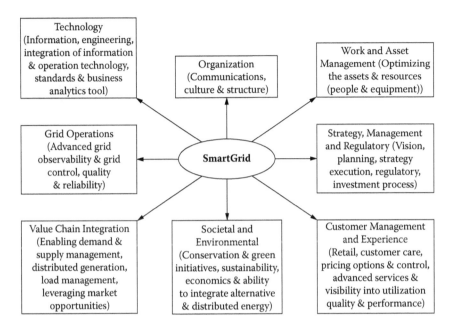

FIGURE 2.5
Vision and logical domain of a smart grid.

spot price. Presently, the independent system operator (ISO) has visibility into the transmission sub-station but generally does not have visibility into the distribution network, where most of the small commercial and the main residential DR take place. Other entities like utility distribution companies (UDCs), load serving entities (LSEs), energy service providers (ESPs) and curtailment service providers (CSPs) interact directly with consumers to bundle the DR and present it to the ISO/RTO (regional transmission organization).

Several methods of load management by DR have been reported. In [260], a model reference adaptive control strategy for interruptible load management that handles load variation is presented. But, the model could not maintain a flat demand response as required. A fully decentralized grid scheduling framework has been reported in [261], but the model is unable to utilize the DR data for optimization of grid operation. Cellular technology-based demand side energy management has been proposed in [262], but the framework is only capable of operating in distribution networks rather than optimizing the whole grid operation. In [263], an optimization algorithm to schedule direct load control DR as a part of a virtual power plant is presented, but the participation of the loads in scheduling the plants is nominal in the cases cited. Harmonic distortion and transformer derating-based load management in the smart grid has been proposed in [264], but the algorithm proposed in this work does not incorporate the cost factors of the DR. Pulse width modulation (PWM)-based direct load scheduling is proposed in [265], but the involvement of distribution networks has not been shown in this work. A DR-based market resource planning strategy has been shown in [266], but the model proposed only shows the DR connectivity in the network rather than enlightening the optimization scheme to be harnessed in utilizing DR for optimal operation of smart grids. A DR-based distribution grid operation model is shown in [267], but its inclusion in the transmission part of the grid has not been shown. An optimal real-time pricing algorithm is depicted in [268] for utility maximization in the smart grid. In this case, the authors could not utilize the DR effectively for load management. In [269, 270], researchers tried to develop different smart grid business models. All these methods are basically based on some developed architectures, which consider DR but cannot maximize the utilization profit in smart grid environments governed by deregulation and distributed generation.

2.3 Concluding Remarks on Existing Efforts

A literature survey reveals that a lot of research work has already been done, and more is still going on, to evaluate and improve the performance of power systems in regulated as well as deregulated environments. Every technique has its own advantages and disadvantages in terms of accuracy, sensitivity, reliability, investment cost, response time and computational complexities. Hence, the choice of the method or the algorithm is solely application specific. In this book, some novel techniques or algorithms for the assessment and amelioration of power system performance in terms of economy, stability and security have been presented. In an endeavour to reconcile the modern power network problems, some remedial methodologies have been developed in this book, which, apart from mitigating constraint violations, produce cost-effective solutions to improve both security and reliability of a network.

Annotating Outline

- The field of improving stability by compensation and operating conditions by optimization is enormously popular among power system engineers, as they can feel hidden opportunities in this regard.
- The scope of improving system performance in terms of stability and efficiency lies in effective utilization of modern technologies like HVDC and FACTS in lieu of installation and maintenance cost.

- The existing network getting transformed from VIS to a deregulated system will work under an optimum stress; hence, new solutions are to be harnessed in the form of compensation, optimization and supervision.
- Issues like cost volatility optimization and congestion management are still to be taken under proper care by efficient formulation of optimization algorithms and their effective implementation in both off-line and on-line monitoring.
- Smart grid technologies are currently undergoing development to modernize power grids to cope with system demand without violating network constraints.

3

Analysis of Voltage Stability of Longitudinal Power Supply System Using an Artificial Neural Network

Accurate and precise estimation of voltage instability has been a great concern of power engineers over the last few decades. The inevitable effects of instability can only be avoided by fast and exact prediction and corresponding preventive action. Classical methods of voltage stability determination have been analyzed to be accurate but not prompt enough for real-time instability computation and prevention. The objective of this chapter is to develop an artificial neural network-based model to predict the voltage instability in real time. The developed model is cost-effective, accurate, sensitive and faster responding with the alteration of the active and reactive power demand.

3.1 Introduction

Voltage control and stability problems are very familiar to the electric utility industry but are now receiving special attention by power system analysts and researchers. With the growing size of a power system along with economic and environmental pressures, the possible threat of voltage instability is becoming increasingly pronounced. Several factors contribute to voltage collapse, such as rise of loading on transmission line, reactive power constraints, on-load tap changer dynamics and load characteristics. Truly, voltage instability implies an uncontrolled decrease in voltage triggered by a disturbance, leading to voltage collapse, and it is primarily caused by system dynamics. It is also believed among professionals that the existing transmission systems will be more and more utilized due to socio-economic concern, which makes it difficult to build new power plants or transmission lines. As a result, the stress on the existing system will increase. There has been ongoing research on the voltage instability in an interconnected system. In this context, it has been found that classical calculations of voltage stability index or point of voltage collapse have been discussed in much of the literature. On the other hand, the artificial neural network (ANN) has attracted a great deal of attention because of its pattern recognition capabilities and its ability to handle corrupted data. ANN has been successfully applied in numerous power engineering problems. It is a massively parallel-distributed information processing unit that has been inspired by the biological nervous system. It reflects a practical classification approach that can draw on the experience and knowledge of an engineer. An ANN is trained with a set of input and output patterns from which it then learns the linking association of input and output. But, its ability to perform well is affected by the chosen training data as well as network topology and training scheme.

The stability indices for recognizing voltage instability or collapse have been depicted in the first part of the following sections. The second part, however, focuses on the artificial neural network technique for pattern recognition. In this respect, the classical calculation of voltage stability indices and implementation of a cost-effective ANN-based model to predict voltage stability to relieve system operators from the tedious calculation have been discussed to establish the applicability of ANN in modern power networks.

3.2 Theoretical Development of Voltage Stability and Voltage Collapse

One of the most fundamental concepts in AC power transmission is stability. It is the property of the power system that enables its operation in the intended mode, where power flows through the entire network and the power angles have their magnitude within specified limits. It maintains synchronism between the synchronous machines (chief sources of power generation) and also ensures that the system voltage and currents do not exceed the rated values. Stability of an AC power system is also denoted by its capability to recover from planned and unplanned electrical disturbances and outages, viz. switching operation, faults, variation in load demand, etc.

Hence, stability studies are an integral part of power system planning and should be carried out in order to ensure appropriate operation of the system. The study of stability is broadly categorized into two parts: angle and voltage stability. Stability, depending upon the disturbance, can be of three different types: steady state, transient and dynamic. The steady-state stability is the ability of a system to return to its normal operation after being subjected to slow and continuous disturbance or increase in load. The transient stability, however, is the capability of a system to restore its operation to a stable point on occurrence of a fast and sudden disturbance like load pull-out. The timescale of dynamic stability lies in between steady-state and transient stability. The voltage stability, however, is the ability of the system to sustain its voltage under variable operating conditions. In the following sub-sections the causes of voltage instability are discussed.

3.2.1 Theoretical Background of Voltage Instability and Its Causes

The voltage decline is often monotonous and small at the onset of the collapse and difficult to detect. A sudden and probably unexpected increase in the voltage decline often marks the start of the actual collapse. It is not easy to distinguish this phenomenon from angle (transient) stability, where voltages can also decrease in a manner similar to voltage collapse (Figure 3.1). Only careful post-disturbance analysis may in those cases reveal the actual cause.

During the last decades there have been one or several large voltage collapses almost every year somewhere in the world. The main reason may be a higher degree of utilization of the power system, leading to a decreasing system security. Also, the changing load characteristics, increased use of air conditioners and electrical heating appliances may be other causes for losing system stability radically. The incidents that lead to a real breakdown of the system are rare, but when they occur they have large repercussions on society.

Due to the continuous increase in reactive power demand, power systems will be used with a smaller margin to voltage collapse. The reasons are the need to use the invested capital efficiently and difficulties in supervising a deregulated market. In this view, voltage stability is believed to be of greater concern in the future, and if not proactively analyzed and controlled, it can be incredibly erratic, causing failure of synchronism of the different parts of the network.

Nearly all types of contingencies and even slowly increased loads could cause a voltage stability problem. The timescale for the course of events,

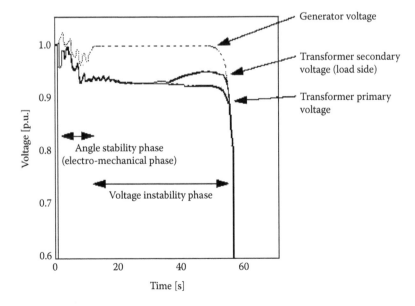

FIGURE 3.1
Example of a collapse simulation of voltage.

which develops into a collapse, may vary from around a second to several tens of minutes. This makes voltage collapse difficult to analyze since there are many phenomena that interact during such a time span (Figure 3.2). Important processes that interact during a voltage decline lasting several minutes are, among others, generation limitation, behaviour of on-load tap changers and load alteration. The actions of these components are often studied in long-term voltage stability studies.

Voltage instability is a very frequent problem that can arise in a power system. A typical property of voltage stability is that the system frequency usually is fairly constant until the very end of its collapse. This indicates that the balance is kept between production and active load demand. Power oscillations between different areas in the system can be a limiting phenomenon on their own, but they may also appear during voltage instability, mixing voltage instability issues with electro-mechanical oscillations.

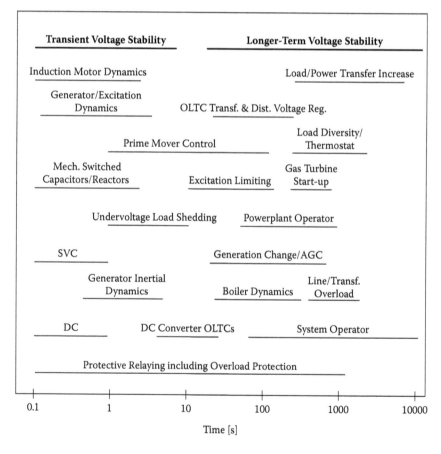

FIGURE 3.2
Different time responses for voltage stability phenomenon.

3.2.2 Few Relevant Analytical Methods and Indices for Voltage Stability Assessment

The typical quasi-steady-state description of a power system applicable to voltage stability analysis is given by the differential algebraic equations:

$$\left.\begin{array}{l} \dot{x} = f(x,y,\lambda) \\ 0 = g(x,y,\lambda) \end{array}\right\} \tag{3.1}$$

where x corresponds to the system state variables and y represents the algebraic variables. The variable λ stands for a parameter or a set of parameters that change slowly with time so that the system moves from one equilibrium point to another until reaching the collapse point. Another way of representing the system is by defining $z = [x, y]^T$, so that (3.1) can be rewritten as

$$\left[\begin{array}{c} \dot{x} \\ 0 \end{array}\right] = F(z,\lambda) \tag{3.2}$$

If we assume that the Jacobian $D_y g(.)[\partial g/\partial y]$ in (3.1) is non-singular along some system trajectories of interest, these equations can be reduced to

$$\dot{x} = f\left(x, y^{-1}(x,\lambda),\lambda\right) = s(x,\lambda) \tag{3.3}$$

This reduction requires that $D_y g(.)[\partial g/\partial y]$ is locally non-singular along these trajectories.

The equilibrium point (z_0, λ_0) of (3.1) is defined by $F(z_0, \lambda_0) = 0$. Hence, based on the non-singularity assumption of the algebraic equations, an equilibrium point (z_*, λ_*), where $D_z F(z_*, \lambda_*)$ (or, $\partial F(z_*, \lambda_*)/\partial z$) is singular, is mathematically known as a singular bifurcation point [271]. This equilibrium point in power systems has been directly related to the voltage collapse problem. Thus, in power systems one is usually interested in determining the singularity of the Jacobian associated with the system dynamic equations. Different models of the system element, particularly generators and loads, affect the location of these collapse points [272, 273]. In addition to that, by changing various parameters in the system one can produce different types of bifurcating phenomena [274].

A typical power flow model of a set of non-linear equations defining the active and reactive power mismatches at system buses is used here to obtain and compare different voltage stability indices as given below:

$$\left[\begin{array}{c} P(u,\lambda) \\ Q(u,\lambda) \end{array}\right] = F(u,\lambda) = 0 \tag{3.4}$$

where $F(u, \lambda)$ is a subset of $F(z, \lambda)$, with u typically representing V and δ, i.e. the magnitude and angle of system bus voltages. In this particular power flow or load flow model of a power system the variations of constant active and reactive power at system buses are assumed to be the main parameter driving the system toward a singularity (or voltage collapse). Although this simple system model is certainly not adequate to thoroughly study the voltage collapse phenomenon, for certain particular dynamic models, the load flow equations yielding adequate results as singularities in the related load flow Jacobian can be associated with actual singular bifurcations of the corresponding dynamical system [271]. Moreover, regardless of the direct relation between singularities of the load flow Jacobian and actual bifurcations of the full dynamical system, it is always of interest to determine system conditions where the load flow problem is not solvable. However, with the help of numerical techniques currently available, (3.4) can be used to compute other system variables besides V and δ so that system controls and their limits may be readily handled by swapping variables in u without the need for changing the structure and number of the equations used in the computational process. For example, generator reactive power injection Q can be part of u, including reactive power mismatch equations at PV buses in (3.4), so that when a Q limit is reached, or released, the corresponding bus voltage magnitude V is swapped for Q in u, or vice versa. Automatic transformer taps or any other control variables can be handled in a similar fashion. It may be noted that $F(u, \lambda)$, and u for that matter, could be modified to include more detailed models of certain system devices such as generators (including automatic voltage regulators (AVRs) and other controls) and loads, or other devices such as high-voltage direct current (HVDC) links and flexible AC transmission system (FACTS), making power flow equations more accurate for computing equilibrium points of the full system model represented by the non-linear function $F(u, \lambda)$.

The λ variable typically represents a scalar parameter or loading factor used to simulate the system load changes that drive the system to collapse in the following way:

$$P_{load_j} = P_{jp}\left(1+k_P\lambda\right)+P_{ji}\left(\frac{V}{V_0}\right)\left(1+k_{Vp}\lambda\right)+P_{jz}\left(\frac{V}{V_0}\right)^2\left(1+k_{Zp}\lambda\right)$$

$$Q_{load_j} = Q_{jQ}\left(1+k_Q\lambda\right)+Q_{ji}\left(\frac{V}{V_0}\right)\left(1+k_{VQ}\lambda\right)+Q_{jz}\left(\frac{V}{V_0}\right)^2\left(1+k_{ZQ}\lambda\right)$$

(3.5)

where P_{load_j} and Q_{load_j} represent the active and reactive load at bus j. $P_{jp}, P_{ji}, P_{jz}, k_P, k_{Vp}, k_{Zp}, Q_{jQ}, Q_{ji}, Q_{jz}, k_Q, k_{VQ}, k_{ZQ}$ and V_0 are all predefined constants that determine the composition of constant power, constant current and constant impedance load [272]. The analytical derivations in this book are derived assuming constant power loads in PQ buses and hence we can write for the jth load bus,

$$P_{load_j} = P_{jp}\left(1+k_P\lambda\right)$$

$$Q_{load_j} = Q_{jQ}\left(1+k_Q\lambda\right)$$

(3.6)

For constant power loads λ represents a net MVA change in the total system load.

3.2.2.1 The PV and VQ Curves for the Small System

The active power-voltage function for the small system has a characteristic form usually called the PV curve (Figure 3.3). As can be seen, there is a maximum amount of power that can be transmitted by the system. Another property of the system is that a specific power can be transmitted at two different voltage levels. The high-voltage/low-current solution is the normal working mode for a power system due to lower transmission losses. One way to write the equations describing this power-voltage relation is

$$V = \sqrt{\alpha \pm \sqrt{\alpha^2 - \beta}} \qquad (3.7)$$

$$\alpha = \frac{E^2}{2} - RP - XQ \quad \text{and} \quad \beta = (P^2 + Q^2)\, Z^2 \qquad (3.8)$$

where E is the sending end voltage, V is the receiving end voltage and R, X are the line resistance and reactance, respectively.

The point of maximum loadability (PML) (maximum power transfer capability) is indicated in Figure 3.3. This point can be calculated either by solving PML from the relation $\alpha^2 = \beta$ from (3.7), by implicit derivation of $dP/dV = 0$ in (3.7) or by evaluating the load flow Jacobian singularity.

Another possibility to demonstrate the capacity of the small system is to show the VQ relation. The necessary amount of reactive power in the load end for a desired voltage level V is plotted in Figure 3.4.

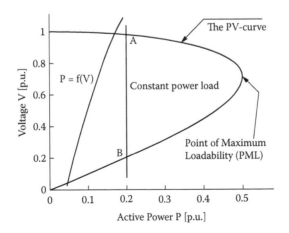

FIGURE 3.3
The PV curve with different load characteristics.

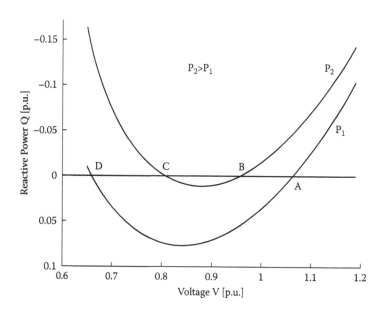

FIGURE 3.4
The VQ curve with two different active loads.

3.2.2.2 Singular Values

Singular values have been employed in power systems because of the useful orthonormal decomposition of the Jacobian matrices. For the real $n \times n$ square Jacobian matrix, $[J] = D_z F (z_0, \lambda_0) \left(\text{or,} \ \dfrac{\partial F(z_0, \lambda_0)}{\partial z} \right)$ at the equilibrium point (z_0, λ_0) of (3.2), we have

$$[J] = R\Sigma S^T = \sum_{i=1}^{n} r_i \sigma_i s_i^T \tag{3.9}$$

where the singular vectors r_i and s_i are the ith columns of the unitary matrices R and S, and Σ is a diagonal matrix of positive real singular values σ_i such that $\sigma_1 \geq \sigma_2 \geq \cdots \geq \sigma_n$. The diagonal entries of Σ^2 correspond to the eigenvalues of matrix JJ^T.

This singular value decomposition is typically used to determine the rank of a matrix, which is equal to the number of non-zero singular values of J. Hence, its application to voltage collapse analysis focuses on monitoring the smallest singular value up to the point when it becomes zero at the collapse point [275].

In general the Jacobian $[J]$ contains the first derivatives of the reactive power mismatch equations $Q(z, \lambda)$ with respect to the voltage magnitude.

Hence, linearizing the steady-state equations $F(z, \lambda) = 0$ at equilibrium point (z_0, λ_0),

$$[\Delta F (z, \lambda)] = [J] [\Delta z] \tag{3.10}$$

or

$$\begin{bmatrix} \Delta P(z, V, \lambda) \\ \Delta Q(z, V, \lambda) \end{bmatrix} = \begin{bmatrix} \partial P(z_0, \lambda_0)/\partial z & \partial P(z_0, \lambda_0)/\partial V \\ \partial Q(z_0, \lambda_0)/\partial z & \partial Q(z_0, \lambda_0)/\partial V \end{bmatrix} \begin{bmatrix} \Delta z \\ \Delta V \end{bmatrix} \tag{3.11}$$

$$\therefore \begin{bmatrix} \Delta P(z, V, \lambda) \\ \Delta Q(z, V, \lambda) \end{bmatrix} = \begin{bmatrix} J_1 & J_2 \\ J_3 & J_4 \end{bmatrix} \begin{bmatrix} \Delta z \\ \Delta V \end{bmatrix} \tag{3.12}$$

For the typical load flow model, $\Delta P (z, V, \lambda)$ represents the active power mismatches, $\Delta Q (z, V, \lambda)$ represents the reactive power mismatches and z represents the bus voltage angles δ. From (3.12) it can be valid for any equilibrium point other than the voltage collapse point, as

$$\begin{bmatrix} \Delta z \\ \Delta V \end{bmatrix} = [J]^{-1} \begin{bmatrix} \Delta P(z, V, \lambda) \\ \Delta Q(z, V, \lambda) \end{bmatrix}$$

$$\begin{bmatrix} \Delta z \\ \Delta V \end{bmatrix} = \sum_{i=1}^{n} \sigma_i^{-1} s_i r_i^T \begin{bmatrix} \Delta P(z, V, \lambda) \\ \Delta Q(z, V, \lambda) \end{bmatrix} \left[\because [J] = \sum_{i=1}^{n} r_i \sigma_i s_i^T \right] \tag{3.13}$$

It may be noted that a minimum singular value is a relative measure of how close the system is to the voltage collapse or singular point. Furthermore, near this collapse point, since σ_n is close to zero, (3.13) can be rewritten as

$$\begin{bmatrix} \Delta z \\ \Delta V \end{bmatrix} = \sigma_n^{-1} s_n r_n^T \begin{bmatrix} \Delta P(z, V, \lambda) \\ \Delta Q(z, V, \lambda) \end{bmatrix} \tag{3.14}$$

Hence, the associated left and right singular vectors, r_n and s_n, contain important information. The maximum entries in s_n indicate the most sensitive voltage magnitudes (critical buses or weak buses), and the maximum entries in r_n correspond to the most sensitive direction for changes of power injections.

Assuming that ΔP $(z, V, \lambda)=0$, which is the standard power flow model corresponding to only reactive power injection changes, (3.13) yields

$$\left[\Delta Q(z, V, \lambda)\right] = \left[\left(J_4 - J_3 J_1^{-1} J_2\right)\right][\Delta V] = \left[J_{QV}\right][\Delta V] \tag{3.15}$$

At voltage collapse point J_1 is not singular even though $[J]$ is singular. Thus, $[J_{QV}]$ becomes singular at the collapse point since

$$\left|J_{QV}\right| = |J| / |J_1| \tag{3.16}$$

where $|J|$ is the determinant of $[J]$; similarly, $|J_1|$ and $|J_{QV}|$ are the determinants of $[J_1]$ and $[J_{QV}]$, respectively. The singular values of this reduced matrix $[J_{QV}]$ can then be used to determine proximity to voltage collapse. Furthermore, these singular values show better profiles than the ones of $[J]$, as has been demonstrated in the literature [276, 277].

It is interesting to highlight the fact that the sub-matrix $[J_3]$ is quasi-symmetric for small values of transmission system resistances. Therefore, one expects a similar attribute for $[J_{QV}]$, making the singular values and eigenvalues for this matrix practically identical [276], as symmetric matrices have similar singular value and eigenvalue decomposition.

3.2.2.3 Eigenvalue Decomposition

Eigenvalues, as singular values, are also often used to determine the proximity to the voltage collapse point [272, 278]. The eigenvalue decomposition for the Jacobian matrix $[J]$ is given by

$$[J] = [\xi] [U] [\eta] \tag{3.17}$$

where $[\xi]$ = right eigenvector of $[J]$, $[\eta]$ = left eigenvector of $[J]$ and $[U]$ = diagonal eigenvalue matrix of $[J]$.

$[J_{QV}]$, as defined in (3.16), is quasi-symmetric and therefore diagonalizable. This decomposition may be applied directly to assess voltage stability [276]. In addition to that, due to its quasi-symmetric structure, a set of only real eigenvalues and eigenvectors can be expected that are very similar in value to the corresponding singular values and singular vectors. Thus, for $[J_{QV}]$, the eigenvectors associated with eigenvalues closest to zero have the same interpretation as the singular vectors near the collapse point. Therefore, the maximum entries in the right eigenvector correspond to the critical buses (most sensitive buses) in the system, and the maximum entries in the left eigenvector pinpoint the most sensitive direction for change of power injections [279, 280]. Comparing the singular value and eigenvalue-based indices to the sensitivity factors, somewhat similar information can be obtained with these indices, but at higher computational cost than in the case of sensitivity factors.

3.2.2.4 Modal Analysis

The static voltage stability analysis is based on the modal analysis of the power flow Jacobian matrix that can be rewritten as follows:

$$
\begin{bmatrix} \Delta P_{PQ,PV} \\ \Delta Q_{PQ} \end{bmatrix} = \begin{bmatrix} J_{P\theta} & J_{PV} \\ J_{Q\theta} & J_{QV} \end{bmatrix} \begin{bmatrix} \Delta\theta \\ \Delta V_{PQ} \end{bmatrix}
\tag{3.18}
$$

where $\Delta P_{PQ,PV}$ is the incremental changes in active power of PQ and PV buses, ΔQ_{PQ} is the incremental changes in reactive power of PQ buses, $\Delta\theta$ is a vector that contains the incremental changes in bus voltage angle and ΔV is a vector that contains the incremental changes in bus voltage magnitude.

The elements of the Jacobian matrix represent the sensitivities between power flow bus voltage changes. According to the classical static voltage stability analysis, power system voltage stability is largely affected by the reactive power. Therefore, keeping active power constant at each operating point, the QV analysis can be carried out.

Assuming $\Delta P_{PQ,PV} = 0$, it follows from (3.18) that

$$
\Delta Q_{PQ} = \left[J_{QV} - J_{Q\theta} \cdot J_{P\theta}^{-1} \cdot J_{PV} \right] \cdot \Delta V_{PQ} = J_R \cdot \Delta V_{PQ}
\tag{3.19}
$$

and

$$
\Delta V_{PQ} = J_R^{-1} \cdot \Delta Q_{PQ}
\tag{3.20}
$$

Based on the J_R^{-1}, which is a reduced VQ Jacobian matrix, the QV modal analysis can be carried out. Therefore, the bus, branch and generator participation factors on the static voltage stability can also be obtained. Moreover, the stability margin and the shortest distance to instability will be determined.

3.2.2.5 Voltage Stability Index L

For a system where n is the total number of buses, with 1, 2, ..., g generator buses, $g + 1, g + 2, ..., g + s$ static VAR compensator (SVC) buses, and $g + s + 1$, ..., n remaining buses ($r = n - g - s$), and t is the number of on-load tap changing (OLTC) transformers, a load flow result can be evaluated for a given operating condition. The load flow algorithm incorporates load characteristics and generator control characteristics. Using the load flow result, the L index [281] can be computed as

$$L_j = \left| 1 - \sum_{i=1}^{i=g} F_{ji} \frac{V_i}{V_j} \right| \qquad (3.21)$$

where $j = g + 1, \ldots, n$ and all the terms within sigma on the right-hand side of (3.21) are complex quantities. The values F_{ji} are obtained from the Y bus matrix as follows:

$$\begin{bmatrix} I_G \\ I_L \end{bmatrix} = \begin{bmatrix} Y_{GG} & Y_{GL} \\ Y_{LG} & Y_{LL} \end{bmatrix} \begin{bmatrix} V_G \\ V_L \end{bmatrix} \qquad (3.22)$$

where I_G, I_L, V_G, V_L, represent current and voltages at the generator nodes and load nodes. Rearranging (3.22), it can be written as

$$\begin{bmatrix} V_L \\ I_G \end{bmatrix} = \begin{bmatrix} Z_{LL} & F_{LG} \\ K_{GL} & Y_{GG} \end{bmatrix} \begin{bmatrix} I_L \\ V_G \end{bmatrix} \qquad (3.23)$$

where $F_{LG} = [Y_{LL}]^{-1} [Y_{LG}]$ are the required values. The L indices for a given load condition are computed for all load buses.

The equation for the L index for the node can be written as

$$L_j = \left| 1 - \sum_{i=1}^{i=g} F_{ji} \frac{V_i}{V_j} \angle \theta_{ji} + \delta_i - \delta_j \right| \qquad (3.24)$$

where δ_i and δ_j are the voltage angles of the ith and jth buses with regard to the slack bus and θ_{ij} is the power factor angle.

For stability, the index L_j must not be violated for any of the nodes j. Hence, the global indicator L describing the stability of the complete sub-system given by L is the maximum of L_j for all j (load nodes). The indicator L is a quantitative measure for the estimation of the distance of the actual state of the system to the stability limit. The local indicators permit the determination of those nodes from which a collapse may originate. The stability margin in this case is obtained as the distance of L from a unit value, i.e. $(1 - L)$. The advantage of this method lies in the simplicity of the numerical calculation and expressiveness of the results.

3.2.2.6 Fast Voltage Stability Index (FVSI) and Line Quality Factor (LQF)

The fast voltage stability index and line quality factor [51] are quite efficient in predicting the voltage collapse of a power network. But their proper

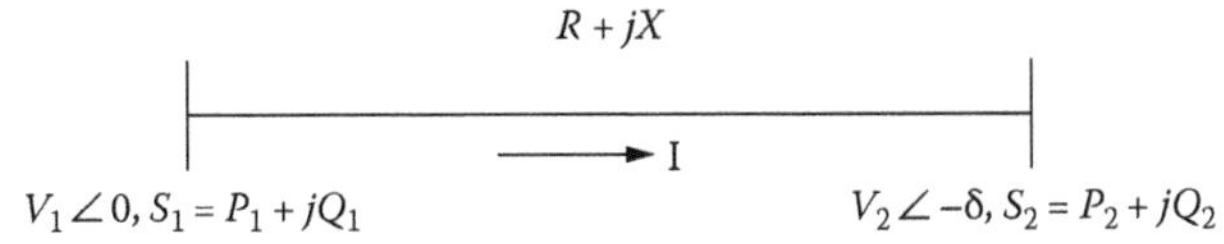

FIGURE 3.5
Typical single-line diagram of transmission line.

formulation is quite imperative, as that will determine the dynamic restoration of stability under disturbance. For that purpose, an interconnected system is reduced to a single-line network and applied to assess the overall system stability. Utilizing the same concept, a criterion for stability can be developed. Let us consider a single line of an interconnected network as shown in Figure 3.5.

The sending end bus voltage (V_1) is actually the summation of line drop and receiving end bus voltage (V_2).

$$V_2\angle-\delta+\left[\frac{P_2-jQ_2}{V_2\angle+\delta}\right](R+jX)=V_1\angle0$$

$$V_2^2+P_2R+Q_2X+j(XP_2-Q_2R)=V_1V_2\cos\delta+jV_1V_2\sin\delta \qquad (3.25)$$

After separating the real and imaginary parts and eliminating δ, (3.25) reduces to

$$V_2^4+V_2^2\left(2Q_2X-V_1^2\right)+X^2\left(P_2^2+Q_2^2\right)=0 \qquad (3.26)$$

To have real solutions for voltage, (3.25) must have real roots. Thus, the following conditions, which can be used as stability criteria, need to be satisfied:

$$\left(2Q_2X-V_1^2\right)^2-4X^2\left(P_2^2+Q_2^2\right)\geq0$$

$$LQF=\frac{4X\left[Q_2+\dfrac{XP_2^2}{V_1^2}\right]}{V_1^2}\leq1.00 \qquad (3.27)$$

Again from (3.25), substituting the value of P_2,

$$V_2^2+Q_2X+V_1V_2\left[\frac{\sin\delta}{\tan\theta}-\cos\delta\right]=0 \qquad (3.28)$$

where $Z\sin\theta=X$

For getting a real solution of voltage, (3.28) should have real roots, and at the limiting condition, the following criterion must be satisfied:

$$FVSI = \frac{4XQ_2}{V_1^2\left[\sin(\delta-\theta)\right]^2} \le 1.00 \tag{3.29}$$

The stability criteria (3.27) and (3.29) are used to find the stability index for each line connected between two bus bars in an interconnected power network. Based on the stability indices of the lines, voltage collapse can be accurately predicted. As long as the stability indices are less than 1, the system is stable.

3.2.2.7 Global Voltage Stability Indicator

For a two bus system, the sending end and receiving end active and reactive powers are represented as P_S, Q_S and P_R, Q_R, respectively, and the power flow equations can be represented as

$$P_S = P_L + P_R \tag{3.30}$$

$$Q_S = Q_L + Q_R \tag{3.31}$$

The active and reactive power losses are given by

$$P_L = \frac{R\left(P_S^2 + Q_S^2\right)}{V^2} \quad \text{and} \quad Q_L = \frac{X\left(P_S^2 + Q_S^2\right)}{V^2} \tag{3.32}$$

where V is the voltage at s the ending end and R and X are the equivalent resistance and reactance of the network.

The power flow equation can be then modified as follows:

$$P_S = \frac{R\left(P_S^2 + Q_S^2\right)}{V^2} + P_R \tag{3.33}$$

$$Q_S = \frac{X\left(P_S^2 + Q_S^2\right)}{V^2} + Q_R \tag{3.34}$$

On simplification, the sending end active power expression becomes

$$P_S = P_R + \frac{R\left\{\dfrac{P_S^2\left(R^2+X^2\right)-\left(2X^2P_R-2XRQ_R\right)P_S+\left(X^2+P_R^2-2XRP_RQ_R+R^2+Q_R^2\right)}{R^2}\right\}}{V^2} \tag{3.35}$$

To have a real solution of active power, the discriminant part must be greater than or equal to zero. This gives the following condition:

$$4\left\{\left(XP_R - RQ_R\right)^2 + XQ_R + RP_R\right\} \leq 1 \qquad (3.36)$$

This term has been named global *VSI* [282], and to maintain the global stability, the criterion is *VSI* ≤ 1.

3.2.2.8 Voltage Collapse Proximity Indicator (VCPI)

With the aid of Thevenin's theorem, a general conclusion can be drawn about the condition for maximum power transfer to a node in a system. The maximum power transfer to a bus takes place when the load impedance becomes the driving point impedance, as seen from the load bus under consideration. Any network of linear elements and energy sources can be represented by a series combination of an ideal voltage and impedance. For a network, the Thevenin's equivalent impedance looking into the port between bus *I* and the ground is Z_{ii}. Therefore, at load bus *i*, with load impedance Z_i, for permissible power transfer to the load at bus *i*, $\left|\dfrac{Z_{ii}}{Z_i}\right| \leq 1$. The voltage collapse proximity indicator (VCPI) [283] for all nodes is computed as

$$VCPI_i = \left|\frac{Z_{ii}}{Z_i}\right| \qquad (3.37)$$

The stability margin in this case is obtained as the distance of VCPI from unity. The bus having the maximum value of VCPI is the weakest bus in the system.

3.2.2.9 Proximity Indices of Voltage Collapse

It has been observed that with increase in load at bus *i*, the values of diagonal elements of Jacobian matrices $\partial Q_i/\partial V_i$ and $\partial P_i/\partial \delta_i$ are reduced. Thus, the deviation in value of $\partial Q_i/\partial V_i$ and $\partial P_i/\partial \delta_i$ from its no-load value to the value at any particular loading condition can be used as an index of voltage stability for the load bus *i*. Using these criteria, two voltage stability indices can be defined as follows [283]:

$$I_{pi} = \frac{\left(\dfrac{\partial P_i}{\partial \delta_i}\right) loading\ condition}{\left(\dfrac{\partial P_i}{\partial \delta_i}\right) no\text{-}load\ condition} \qquad (3.38)$$

$$I_{qi} = \frac{\left(\dfrac{\partial Q_i}{\partial V_i}\right) loading\ condition}{\left(\dfrac{\partial Q_i}{\partial V_i}\right) no\text{-}load\ condition} \tag{3.39}$$

3.2.2.10 Identification of Weak Bus of Power Network

The governing equation of load flow analysis used in the N-R method is given by

$$\begin{bmatrix} \Delta P \\ \Delta Q \end{bmatrix} = \begin{bmatrix} \dfrac{\partial P}{\partial \delta} & |V|\dfrac{\partial P}{\partial |V|} \\ \dfrac{\partial Q}{\partial \delta} & |V|\dfrac{\partial Q}{\partial |V|} \end{bmatrix} \begin{bmatrix} \Delta \delta \\ \dfrac{\Delta |V|}{|V|} \end{bmatrix} \tag{3.40}$$

Hence, the real and reactive power sensitivities of the *i*th bus can be written as [284]

$$\frac{\partial P_i}{\partial |V_i|} = \frac{[J_2]_{ii}}{[V_i]}; \qquad \frac{\partial Q_i}{\partial |V_i|} = \frac{[J_4]_{ii}}{[V_i]} \tag{3.41}$$

Equation (3.41) represents the real and reactive power sensitivities of the *i*th bus. $\dfrac{\partial Q_i}{\partial |V_i|}$ also indicates the degree of weakness for the *i*th bus as $\dfrac{\partial Q_i}{\partial |V_i|}$ is high, and $\dfrac{\partial |V_i|}{\partial Q_i}$ becomes low, indicating a minimum change in voltage for variation of Q (reactive power) of the bus. Thus, a higher value of $\dfrac{\partial Q_i}{\partial |V_i|}$ represents a lesser degree of weakness of the *i*th bus. With the help of this indicator, the weakest bus of a power network can be evaluated.

3.2.2.11 Diagonal Element Ratio

This index is given as the ratio of the maximum to minimum values of diagonal elements of the Jacobian matrix [283]. Mathematically, it can be expressed as

$$I_d = \frac{J_{max}}{J_{min}} \tag{3.42}$$

where J_{max} and J_{min} are maximum and minimum values of diagonal elements of the Jacobian matrix.

3.2.2.12 Line Voltage Stability Index

The complex power injected at bus k consists of an active and a reactive component and may be expressed as a function of nodal voltage and the injected current at the bus:

$$S_k = P_k + jQ_k = E_k I_k^* = E_k \left(Y_{kk} E_k + Y_{km} E_m \right)^* \tag{3.43}$$

where I_k^* is the complex conjugate of the current injected at bus k and the other terms carry their usual meanings. The expressions for P_k and Q_k can be determined by using the expressions of E_k, Y_{kk}, Y_{km}:

$$P_k = V_k^2 G_{kk} + V_k V_m \left[G_{km} \cos(\theta_k - \theta_m) + B_{km} \sin(\theta_k - \theta_m) \right] \tag{3.44}$$

$$Q_k = -V_k^2 B_{kk} + V_k V_m \left[G_{km} \sin(\theta_k - \theta_m) - B_{km} \cos(\theta_k - \theta_m) \right] \tag{3.45}$$

With the help of these expressions the Jacobian matrix for load flow analysis can be developed, and the point of voltage instability being coincident with the singularity of the Jacobian, the condition of voltage stability can be obtained from the following condition:

$$\Delta[J] = 0 \tag{3.46}$$

Thus, at the voltage collapse point, by using (3.44) to (3.46) the line voltage stability index can be formulated as

$$LVSI = \frac{V_m}{V_k} \cos \theta_m = \frac{1}{2} \tag{3.47}$$

3.2.2.13 Local Load Margin

This index is based on physical quantities of a power flow model. T. Nagao et al. [285] have proposed the index for bus i as follows:

$$P_{LLM_i} = \frac{P_{\max_i} - P_{initial_i}}{P_{\max_i}} \tag{3.48}$$

where $P_{initial_i}$ is the initial load in MW and $P_{\max_i}$ (in MW) is the load at the nose of the PV curve when the load at bus i is increased at a fixed power factor.

The local load margin P_{LLM_i} has a value between 1 and 0 (at the collapse point). Here the loads at other buses are assumed to remain constant.

However, Equation (3.48) allows for the computation of a voltage stability margin for each load point. As the voltage may be on a lower solution domain depending on the system conditions, it is convenient to present the value P_{LLM_i} as a negative number when the system voltage is in this particular state.

As P_{LLM_i} is defined with respect to a specific bus, its computation is relatively easy. However, a voltage stability margin should be evaluated for the entire power system; thus, the index P_{LLM_i} should be computed for all load buses. This index can be utilized to monitor the worst possible contingencies in the system for the stability point of view and can be used to prepare a ranking table according to the severity of line contingencies. Hence, this index can be employed as a contingency factor in contingency analysis of a power network.

3.2.2.14 Voltage Ratio Index

The V/V_0 index is rather simple to define and compute. Thus, assuming the bus voltage values (V) are known from load flow or state estimation studies, new bus voltages (V_0) are obtained by solving a load flow for the system at an identical state but with all loads set to zero. The ratio V/V_0 at each node yields a voltage stability map of the system, allowing for immediate detection of weak and effective countermeasure spots. A problem with this index is that it presents a highly non-linear profile with respect to changes on the system parameter λ, not allowing for accurate predictions of proximity to collapse [278]. Nevertheless, when used together with operating experience, it has turned out in practice to be an effective tool against voltage collapse. Thus, the V/V_0 index has been successfully used since 1982 in Belgium for off-line studies, particularly in seasonal operation planning; it was then added in 1995 to the on-line security assessment unit of the new Belgium dispatch centre.

3.3 Theory of ANN

Research in the field of neural networks has been attracting increasing attention in recent years. Since 1943, when Warren McCulloch and Walter Pitts presented the first model of artificial neurons, the study of neurons, their interconnections and their role as the brain's elementary building blocks has become one of the most dynamic and important areas of exploration of the modern research fields. The earlier trends followed in the technological fields got enriched after the incorporation of ANN-based models. In quite a few cases, application of ANN could even produce better solutions with respect to the classical techniques available. Apart from that, ANN has projected quite a few drawbacks of classical approaches to increase the adoption of artificial intelligence (AI) in the engineering fields:

1. Classical techniques impose restrictions on the number of input data: one is limited to a few inputs among dozens or hundreds available, imposing a priori variable selection, with all the inherent pitfalls.
2. Regressions are performed using simple dependency functions that are not very realistic. The hypothesis is made that there is only one dependency function over the whole data set, instead of many distinct niches.
3. Other hypotheses imposed by their underlying theories (normal distributions, equi-probabilities, uncorrelated variables) are known to be violated, but are necessary for their good operation.

In all the cases one has to call on an expert of the method and perform many weeks[2], even a few months' worth of work. In comparison to classical techniques, ANN offers the following advantages:

1. Neural nets need less constraining hypotheses (the dependency must be a function, nothing more).
2. Qualitative (enumerative) data are straightforwardly handled.
3. No preprocessing, simplification or reduction of quantitative variables is necessary.
4. A developed ANN-based model, for predicting voltage stability, does not account for any added monetary investment. Hence, such methods of system performance evaluation are very much cost-effective, which will not hinder consumer welfare by an unnecessary increase in electricity price.

It has been demonstrated that linear regression and logistic regression are particular cases of neural nets (with one layer and a linear threshold function). The same thing happens with principal component analysis (PCA), whose values are contained in the weights of the neurons of a one-layer linear threshold function network performing self-association.

3.3.1 Attributes of ANNs

Artificial neural networks are a form of computation inspired by the structure and function of the brain. A common topology used is the weighted directed graph. The nodes, known as neurons, in the graph can be on or off and time is discrete. At each time instant, all the on nodes send an impulse along their outgoing arcs to their neighbouring nodes. All nodes sum their incoming impulses, weighted according to the arc. All the nodes at which this sum exceeds a threshold turn on at the next time instant; all others turn off. Computation proceeds by setting nodes, waiting for the network to reach a steady state, and then reading the output nodes. Nodes can be trained to recognize certain patterns, for instance to classify objects by their features. In this section, building blocks of ANN and different networks have been discussed that are very much imperative for application of ANN in voltage stability analysis of power networks.

3.3.1.1 Building Block of ANNs

First attempts at building artificial neural networks were motivated by the desire to create models for natural brains. Much later it was discovered that ANNs are a very general statistical framework for modeling posterior probabilities given a set of samples (the input data). This simple element consists of the components given below:

1. A set of inputs
2. A set of weights
3. A threshold
4. An activation function
5. A single neuron output or set of outputs

Inputs: Typically, these values are external stimuli from the environment or come from the outputs of other artificial neurons. They can be discrete values from a set, such as {0, 1}, or real-valued numbers.

Weights: These are real-valued numbers that determine the contribution of each input to the neuron's weighted sum and eventually its output. The goal of neural network training algorithms is to determine the best possible set of weight values for the problem under consideration. Finding the optimal set is often a trade-off between computation time and minimizing the network error. Apart from that the weights incorporated depend upon the type of input.

The basic building block of a (artificial) neural network is a neuron shown in Figure 3.6, which is a processing unit with some (usually more than one) inputs and only one output.

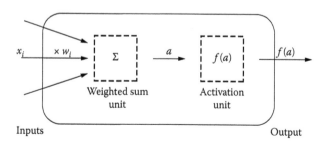

FIGURE 3.6
The neuron.

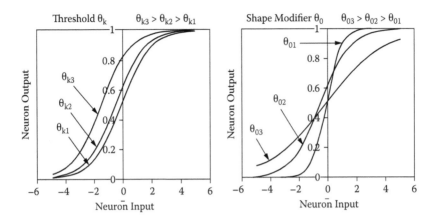

FIGURE 3.7
Threshold and shape modifier used in ANN.

At the outset, each input x_i is weighted by a factor w_i and the whole sum of input is calculated as $\sum_{allinputs} w_i x_i = a$. Then an activation function f is applied to the result a and the neuronal output is taken to be $f(a)$.

Threshold: The threshold is a real number that is added algebraically to the weighted sum of the input values. Sometimes the threshold is referred to as a bias value.

Activation function: The activation function for the original McCulloch–Pitts neuron was the unit step function. However, the artificial neuron model has been expanded to include other functions such as the sigmoid, piecewise linear, Gaussian, etc., the most popular of which is the sigmoidal function.

The effect of threshold and shape modifier can be explained by the figures (Figure 3.7), which are self-expounding.

In accordance with the working environment and inputs, the threshold and shape modifier make the activation function more reliable. This also assists to achieve optimal weight values of the inputs of each neuron.

3.3.1.2 Building Layers of ANNs

Generally ANNs are built by putting the neurons in layers and connecting the outputs of neurons from one layer to the inputs of the neurons from the next layer (Figure 3.8).

The depicted network is also named as feed-forward network (a feed-forward network does not have feedbacks, i.e. no loops). In fact, there is no

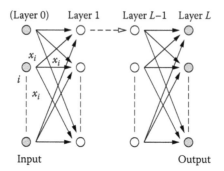

FIGURE 3.8
The general layout of a neural network.

processing on layer 0; its role is just to distribute the inputs to the next layer (data processing really starts with layer 1). For this reason its representation will be omitted most of the time.

Generally, the output of one neuron may go to the input of any neuron, including itself. If the outputs of neurons from one layer are going to the inputs of neurons from previous layers, then the network is called recurrent, this providing feedback. Lateral feedback is done when the output of one neuron goes to the other neurons on the same layer. So, to compute the output, an activation function is applied on the weighted sum of inputs.

$$total\ inputs = a = \sum_{all-inputs} w_i.x_i \qquad (3.49)$$

$$output = activation\ function\left(\sum_{all-inputs} w_i.x_i\right) = f(a) \qquad (3.50)$$

Neural nets have two fundamental advantages:

1. It has the ability to represent any function, which may be linear or not, simple or complicated. Neural nets are what mathematicians call universal approximators (Kolmogorov's theorem, 1957).
2. The ability to learn from representative examples. Model building is automatic.

However, building a neural model belongs to data analysis and not to magic (even though, to quote Arthur C. Clarke, 'sufficiently advanced technology is indistinguishable from magic'). The data must be explicative and in sufficient amount.

3.3.1.3 Structures of Neural Networks

A neural network comprises the neuron and building blocks. The behaviour of the network depends largely on the interaction between these building blocks. There are three types of neuron layers: input, hidden and output layers. Two layers of neurons communicate via a weight connection network. There are different types of weighted connections, some of which are discussed below.

3.3.1.3.1 Feed-Forward Neural Network

For this kind of neural models, data from neurons of a lower layer are propagated forward to neurons of an upper layer via feed-forward connection networks. The feed-forward neural network shown in Figure 3.9 has an input vector X consisting of components x_1, x_2, and x_n, a hidden layer Y having the components of y_1, y_1, and y_m and an output vector Z having the components of z_1, z_2, and z_k. The synaptic links carrying weights connect every input neuron to output neurons. The function of hidden neurons is to intervene between the external input and the network output in some useful manner [286]. By adding one or more hidden layers, the network is enabled to extract higher-order statistics. The set of output signals of the neurons in the output layer of the network constitutes the overall response of the network to the activation pattern supplied by the source nodes in the input layer.

3.3.1.3.2 Recurrent Neural Network

The backpropagation neural networks are strictly feed-forward networks in which there are no feedbacks from the outputs of one layer to the inputs of the same layer or earlier layers of neurons. However, such networks have no memory, since the output at any instant depends entirely on the inputs and the weights at that instant.

There are situations (e.g. when dynamic behaviour is involved) where it is advantageous to use feedback in neural networks. When the output

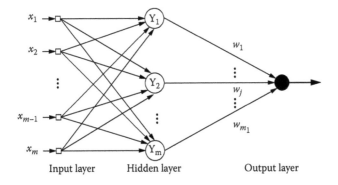

FIGURE 3.9
A feed-forward neural network.

of a neuron is fed back into a neuron in an earlier layer, the output of that neuron is a function of both the input from the previous layer at time t and its own output that existed at an earlier time, that is at time $t - \Delta t$, where Δt is the time for one cycle of calculation. Hence, such a network exhibits characteristics similar to short-term memory, because the output of the network depends on both the correct and prior inputs.

Neural networks that contain such feedback are called recurrent neural networks. Although virtually all the neural networks that contained feedback could be considered recurrent networks, the discussion here will be limited to those that use backpropagation for training (often called recurrent backprop networks). Let us consider an elementary feed-forward network with the input, middle and output layers each having only one neuron, and where the neuron h is a buffer neuron that instantaneously sends the input x to neuron p. When the input $x(0)$ (x at time $t = 0$) is applied to the input, the outputs of neuron p and q at time $t = 0$, $v(0)$ and $y(0)$, respectively, are

$$v(0) = \left\{ \phi\left[W_{ij} x(0) \right] \right\}$$

$$y(0) = \phi\left\{ W_{jk} \right\}\left[v(0) \right] = \phi\left\{ W_{jk} \left\{ \phi\left[W_{ij} x(0) \right] \right\} \right\}$$

(3.51)

where ϕ is the activation function operator (usually a sigmoidal function).

3.3.1.3.3 Elman Backpropagation Neural Network

An Elman backpropagation neural network (EBNN) is a two-layer backpropagation network, with the addition of a feedback connection from the output of the hidden layers to its input, as shown in Figure 3.10. This feedback path allows an Elman network to learn to recognize and generate temporal as well as special patterns.

Since there is a feedback connection from the first-layer output to the first-layer input, a recurrent connection is established, which allows the Elman network to both detect and generate time-varying patterns [287]. The delay in this connection stores values from the previous time step, which can be used in the current time step. Here, a specific group of units called context units receives feedback signals from the previous time step. The weights on the feedback connection to the context units are fixed, and information processing is sequential in time. Therefore, training in this network is more difficult than for a standard backpropagation network.

At time t, the activation of the context units implies the activation of the hidden units at the previous time step. The weights of the context units to the hidden units are trained in exactly the same manner as the weights from the input units to the hidden units. Thus, at any time step the training algorithm is the same as for standard backpropagation.

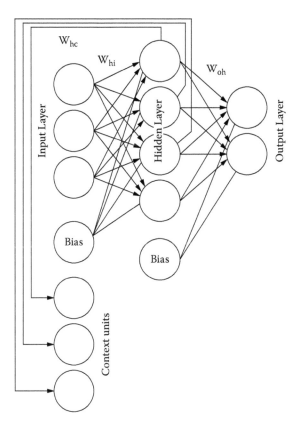

FIGURE 3.10
An Elman backpropagation neural network.

Let W_1 be the weight matrix between the input layer and hidden layer, W_2 be the weight matrix between the hidden layer and output layer and W_3 be the weight matrix between the context layer and hidden layer. i, h, o and c indicate the number of input nodes, hidden nodes, output nodes and context layer nodes. In terms of weight matrix in the neural structure, each weight coefficient can be defined as an element of these matrices, e.g. $W_{1hi} \in W_1$, $W_{2oh} \in W_2$ and $W_{3hc} \in W_3$.

The outputs of the neuron in the hidden and output layers are given as

$$Y_h^{(s)} = f\left(\sum_{i=1}^{n} W_{1hi} x_i^{(s)} + \sum_{h=1}^{m} W_{3hc} y_h^{(s-1)} \right) \tag{3.52}$$

$$Z_o^{(s)} = f\left(\sum_{h=1}^{m} W_{2oh} Y_h^{(s)} \right) \tag{3.53}$$

An updated weight coefficient can be assigned to minimize the approximation error E in the output layers as follows:

$$W_{new} = W_{old} + \eta \Delta W \qquad (3.54)$$

where η is the learning rate,

$$E(W) = (1/2)\sum_{s=1}^{p}\sum_{o=1}^{c}[T_o^{(s)} - Z_o^{(s)}] \qquad (3.55)$$

where $T_o^{(s)}$ is the target value, p is the length of training sequence and W_1 and W_3 are the target coefficient matrices.

$$\Delta W_{3hc} = \sum_{o=1}^{c}[T_o^{(s)} - Z_o^{(s)}]W_{2oh}f\left(Y_h^{(s)}Y_h^{(s-1)}\right) \qquad (3.56)$$

This value is used to update the weight coefficients between the context layer and the hidden layer of the training procedure of the Elman backpropagation neural network.

3.3.1.3.4 Input Delay Feed-Forward Backpropagation Neural Networks

The input delay feed-forward backpropagation neural network is a time delay neural network whose hidden neurons and output neurons are replicated across time. As shown in Figure 3.11, the delay is taken from the top of the input nodes.

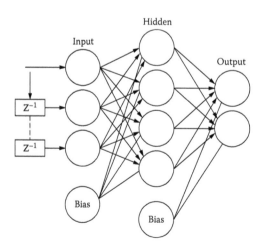

FIGURE 3.11
Input delay feed-forward backpropagation neural network.

Hence, here the network has tapped a delay line that senses the current signal and previous signal, and the delayed signal before it is connected to the network weight matrix through delay time units such as 0, 1 and 2. These are added in ascending order from left to right to correspond to the weight matrix. All other features are similar to the feed-forward neural network, except that the input given to the network is delayed. In this network, memory is limited by the length of the tapped delay line.

Let us consider a non-uniform sampling

$$X_i(t) = X(t - W_i) \tag{3.57}$$

where W_i is the integer delay associated with the component i. Each input is really a convolution of the original input sequence,

$$X_i(t) = \sum_{l=1}^{t} C_i(t-l)l \tag{3.58}$$

In case of delay line memories,

$$C_i(t) = \begin{cases} 1 & t = W_i \\ 0 & otherwise \end{cases} \tag{3.59}$$

3.3.1.4 Training Algorithms of Neural Networks

The above discussed neural networks are commonly categorized in terms of their corresponding training algorithms: fixed-weight networks, unsupervised networks and supervised networks. There is no learning required for the fixed-weight networks (e.g. Hamming net) [287]. In other networks, the learning mode is either unsupervised or supervised.

3.3.1.4.1 Unsupervised Learning

For an unsupervised learning rule, the training set consists of input training patterns only. Therefore, the network is trained without the benefit of any teacher. The network learns to adapt, based on the experiences collected through the previous training patterns. A typical unsupervised system is Kohonen's self-organizing feature map.

3.3.1.4.2 Supervised Learning

Supervised learning networks represent the main stream of the development in neural networks. The convergence property and accuracy of the training process are heavily dependent upon the scaling of the input-output data.

In supervised training, the training patterns must be provided in input-teacher pattern pairs. Depending on the nature of the teacher's information, there are two approaches to supervised learning. One is based on the correctness of the decision, and the other based on the optimization of a training cost criterion. Of the later, the least-square error approximation-based formulation represents the most important special case.

3.3.1.4.3 Phases of Supervised Network

Two phases are involved in a supervised learning network: training phase and retrieving phase.

In the training phase, a training data set is used to determine the weight parameters that define the neural model. This trained neural model will be used later in the retrieving phase to process real test patterns and yield classification results.

Real-world applications may face two very different kinds of real-time processing requirements. One requires real-time retrieving but off-line training speed. The other demands both retrieving and training in real-time. These two lead to very different processing speeds, which in turn affect the algorithm and hardware adopted.

3.4 Analysis of Voltage Stability of Multi-Bus Power Network

Ascertaining voltage stability of a large interconnected network is the primary approach of power system planning. As discussed earlier, monitoring stability and standardizing the same is imperative to insulate the system from sudden and steady-state disturbances. In this section, a comprehensive study on voltage stability surveillance by classical methods has been presented along with a novel ANN-based stability assessment technique. It has been established that the ANN-based analytical tool can provide fairly accurate results compared to the classical method.

3.4.1 Classical Analysis of Voltage Stability

In the classical technique, the foremost step is to determine the weakest bus of the system using the assistance of a standard technique (using (3.41)). Once the weakest bus has been determined, the study concentrates on the effect of the increased load (active or reactive) at that particular bus, and this study is quite capable of revealing a scenario of the whole system stability at stressed conditions. The stability indices have been determined for both the IEEE 30 bus system and a 203 bus-265 line practical power network (eastern grid of India). Tables 3.1 and 3.2 enumerate the summarized description of the aforementioned systems. The single line diagrams (SLDs) have been provided in Appendix A.

TABLE 3.1

System Description of IEEE 30 Bus System

Sl No.	Particulars	Provision
1	Buses	30
2	Branches	41
3	Generators	6
4	Total active demand (MW)	283.6
5	Total reactive power demand (MVAR)	126.2

TABLE 3.2

Description of Eastern Grid of India

Sl No.	Particulars	Provision
1	Buses	203
2	Branches	265
3	Generators	24
4	Total active power demand (MW)	7619
5	Total reactive power demand (MVAR)	4808.35

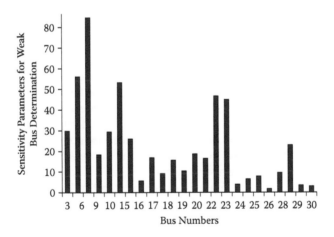

FIGURE 3.12

Weak bus determination of IEEE 30 bus system.

For the IEEE 30 bus system, it has been found that bus number 26 is the weakest one (Figure 3.12), followed by bus numbers 30 and 29. However, bus number 26 has only one interconnecting line with the rest of the system, and an outage of this line will cause complete isolation of bus number 26 from the network. Hence, in the present work the second weakest bus, i.e. bus 30, has been considered and its loading impact on steady-state voltage stability has been observed.

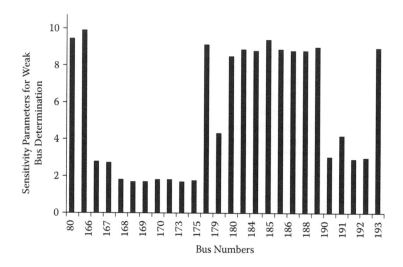

FIGURE 3.13
Weak bus determination of eastern grid of India.

The same procedure has been repeated for the practical network, and as shown in Figure 3.13, bus number 168 is the weakest bus, followed by 169 (weaker bus) and 173 (weak bus) for the considered practical system.

After determining the weakest bus for both the systems, classical analysis has been carried out to study the stability with the continuous increase in reactive demand as voltage magnitude intensely depends on reactive power. A few results are shown in Tables 3.3 and 3.4. These tables also demonstrate the effect of load intensification on voltage magnitude and angle of the weakest bus. The limiting values of the indices have been depicted in Section 3.2.2, and with the increase of the demand the indices are approaching their marginal values. These data sets can be provided to the ANN-based model as training data for the concerned systems.

Thus, through classical analysis different stability indices can be calculated to predict the condition of the system at different loading stresses. In the next part, the work has focused on the realization of ANN in predicting the system instability during different loading conditions. This will assist in avoiding complex classical calculation even in very large practical systems for both on-line and off-line monitoring.

3.4.2 Application of ANN on Voltage Stability Analysis

As discussed earlier, the Elman network-based model can be trained to predict the stability for a particular loading stress. The methodology [306] has been summarized in the flowchart shown in Figure 3.14.

Referring to the flowchart, the immediate step after obtaining stability indices through classical analysis with different loading stresses is the training

TABLE 3.3

Comparison of Different Indices for Voltage Stability Analysis of IEEE 30 Bus System

Reactive Load (p.u.)	L Index	LQF (27–30)	LQF (29–30)	$I_d = J(max)$ $J(min)$	FVSI (27–30)	FVSI (29–30)	I_q (30)
0.019	0.90087	0.04624	0.03614	49.91698	0.07901	0.06174	0.97883
0.025	0.89679	0.06135	0.04801	49.97291	0.10424	0.08166	0.97200
0.030	0.89336	0.07412	0.05808	50.02014	0.12537	0.09841	0.96625
0.035	0.88989	0.08707	0.06832	50.06795	0.14661	0.11532	0.96046
0.045	0.88285	0.11351	0.08931	50.16538	0.18940	0.14960	0.94870
0.055	0.87565	0.14069	0.11102	50.26535	0.23263	0.18454	0.93673
0.065	0.86830	0.16864	0.13348	50.36802	0.27633	0.22019	0.92452
0.075	0.86079	0.19740	0.15674	50.47355	0.32054	0.25660	0.91206
0.085	0.85309	0.22699	0.18085	50.67962	0.36529	0.29381	0.89934
0.095	0.84521	0.25746	0.20586	50.96828	0.41060	0.33189	0.88633
0.105	0.83713	0.28885	0.23182	51.26670	0.45653	0.37091	0.87303
0.120	0.82459	0.33776	0.27271	51.73433	0.52668	0.43135	0.85245
0.130	0.81593	0.37166	0.30138	52.06069	0.57435	0.47304	0.83829
0.140	0.80702	0.40667	0.33127	52.39993	0.62283	0.51598	0.82374
0.150	0.79782	0.44285	0.36248	52.75324	0.67217	0.56028	0.80876
0.160	0.78831	0.48028	0.39514	53.12200	0.72246	0.60607	0.79332
0.170	0.77846	0.51905	0.42937	53.50780	0.77379	0.65352	0.77737
0.180	0.76823	0.55926	0.46533	53.91255	0.82627	0.70280	0.76085
0.190	0.75759	0.60103	0.50322	54.33852	0.88002	0.75415	0.74371
0.200	0.74648	0.64450	0.54324	54.78843	0.93520	0.80783	0.72586
0.210	0.73484	0.68984	0.58568	55.26564	0.99200	0.86417	0.70720
0.220	0.72259	0.73726	0.63087	55.77436	1.05063	0.92358	0.68763
0.230	0.70963	0.78701	0.67924	56.31996	1.11140	0.98660	0.66699
0.240	0.69583	0.83942	0.73133	56.90946	1.17468	1.05391	0.64508
0.250	0.68101	0.89492	0.78790	57.55242	1.24097	1.12644	0.62163
0.260	0.66493	0.95412	0.84998	58.26232	1.31099	1.20552	0.59628
0.270	0.64721	1.01788	0.91912	59.05929	1.38577	1.29312	0.56845
0.280	0.62726	1.08757	0.99777	59.97572	1.46701	1.39245	0.5372
0.290	0.60397	1.16561	1.09038	61.06971	1.55772	1.50934	0.50111
0.300	0.57488	1.25730	1.20686	62.48620	1.66470	1.65698	0.45595
0.310	0.53058	1.38212	1.38434	71.72044	1.81345	1.88562	0.38823
0.311	0.52366	1.39978	1.41192	73.57086	1.83509	1.92179	0.37796
0.312	0.51537	1.42021	1.44484	75.89616	1.86041	1.96527	0.36609
0.313	0.50427	1.44630	1.48870	79.72423	1.89331	2.02380	0.34809
0.314	0.46881	1.52120	1.62857	87.19025	1.99219	2.21510	0.31763
0.315	0.29749	1.94628	2.63750	87.79881	2.56525	3.61075	0.31538
0.316	0.02910	2.42436	4.00148	22744.45	218.838	388.849	0.02286

TABLE 3.4

Comparison of Different Indices for Voltage Stability Analysis of Eastern Grid of India

Bus No.	Reactive Demand (p.u.)	L Index	LQF	Bus Voltage Mag. (p.u.)	Angle (rad)
168	0.100	0.8685	0.2639	0.9976	−0.4176
	0.145	0.8449	0.3443	0.9723	−0.4216
	0.175	0.8284	0.3980	0.9546	−0.4245
	0.200	0.8141	0.4427	0.9392	−0.4271
	0.235	0.7931	0.5054	0.9166	−0.4311
	0.275	0.7673	0.5771	0.8890	−0.4363
	0.300	0.7501	0.6220	0.8705	−0.4399
	0.326	0.7311	0.6687	0.8502	−0.4441
	0.350	0.7124	0.7119	0.8301	−0.4484
	0.376	0.6904	0.7587	0.8066	−0.4537
	0.410	0.6583	0.8201	0.7722	−0.4621
	0.450	0.6126	0.8925	0.7235	−0.4754
	0.476	0.5743	0.9399	0.6827	−0.4880
	0.507	0.4994	0.9970	0.6034	−0.5174
	0.509	0.5004	1.0008	0.5939	−0.5215
169	0.100	0.8215	0.2697	0.9741	−0.4832
	0.150	0.7948	0.3631	0.9448	−0.4880
	0.175	0.7807	0.4098	0.9294	−0.4906
	0.200	0.7660	0.4566	0.9133	−0.4934
	0.225	0.7507	0.5035	0.8965	−0.4965
	0.250	0.7346	0.5505	0.8789	−0.4999
	0.275	0.7177	0.5975	0.8604	−0.5036
	0.300	0.6997	0.6446	0.8407	−0.5076
	0.340	0.6682	0.7202	0.8063	−0.5153
	0.380	0.6321	0.7960	0.7669	−0.5248
	0.400	0.6115	0.8341	0.7444	−0.5308
	0.430	0.5754	0.8916	0.7052	−0.5420
	0.450	0.5457	0.9301	0.6730	−0.5523
	0.473	0.4984	0.9750	0.6219	−0.5708
	0.486	0.4474	1.0014	0.5670	−0.5943
173	0.100	0.8147	0.2940	0.9811	−0.5116
	0.140	0.7932	0.3674	0.9575	−0.5160
	0.180	0.7704	0.4409	0.9326	−0.5209
	0.210	0.7523	0.4963	0.9128	−0.5249
	0.240	0.7331	0.5518	0.8918	−0.5294
	0.270	0.7127	0.6074	0.8695	−0.5344
	0.290	0.6982	0.6446	0.8538	−0.5381
	0.310	0.6830	0.6819	0.8372	−0.5422
	0.350	0.6496	0.7568	0.8008	−0.5517

TABLE 3.4 (*Continued*)

Comparison of Different Indices for Voltage Stability Analysis of Eastern Grid of India

Bus No.	Reactive Demand (p.u.)	L Index	LQF	Bus Voltage	
				Mag. (p.u.)	Angle (rad)
173	0.375	0.6259	0.8038	0.7752	–0.5587
	0.400	0.5993	0.8512	0.7463	–0.5677
	0.425	0.5679	0.8989	0.7124	–0.5790
	0.440	0.5455	0.9278	0.6882	–0.5877
	0.463	0.4999	0.9729	0.6393	–0.6074
	0.477	0.4498	1.0017	0.5858	–0.6328

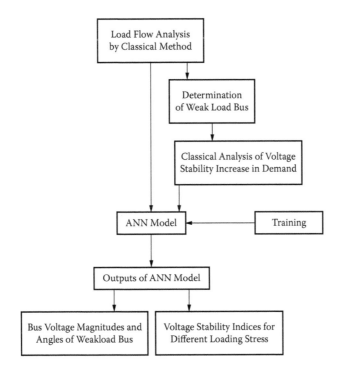

FIGURE 3.14
Developed solution methodology.

procedure of the ANN-based model with the same data set of input training vectors to meet convergence criteria. The convergence criterion for the application has been achieved in 5000 iterations with a training goal of 0.0001. All the other parameters of the described model have been tabulated in Table 3.5.

After achieving the training goal, the network gets fully trained with the correlation between the changing load pattern and the corresponding values of voltage magnitudes, angles and stability indices. The generalizing

TABLE 3.5

Design Parameters of Developed Neural Network

Parameter	ANN Application
Training vector	200
Testing vector	50
Input neurons	2
Hidden layers	1
Output neurons	4
Neurons in hidden layer	100

TABLE 3.6

Comparison of Results Obtained by Classical Calculation and ANN-Based Model (for IEEE 30 Bus System)

Results Obtained from Classical Calculation				Results Obtained from ANN Algorithm			
FVSI (27–30)	FVSI (29–30)	LQF (27–30)	LQF (29–30)	FVSI (27–30)	FVSI (29–30)	LQF (27–30)	LQF (29–30)
0.0872	0.0650	0.0487	0.0381	0.0874	0.0652	0.0488	0.0375
0.1296	0.1017	0.0767	0.0601	0.1294	0.0970	0.0756	0.0614
0.1466	0.1153	0.0871	0.0683	0.1465	0.1154	0.0860	0.0694
0.1894	0.1496	0.1135	0.0893	0.1896	0.1488	0.1132	0.0903
0.1980	0.1565	0.1189	0.0936	0.1974	0.1562	0.1181	0.0952
0.2153	0.1705	0.1297	0.1022	0.2168	0.1706	0.1302	0.1020
0.2413	0.1916	0.1462	0.1155	0.2428	0.1917	0.1467	0.1152
0.2588	0.2058	0.1574	0.1244	0.2583	0.2054	0.1566	0.1252
0.3295	0.2640	0.2032	0.1615	0.3290	0.2634	0.2024	0.1634
0.4015	0.3242	0.2513	0.2008	0.4018	0.3243	0.2510	0.2002
0.4565	0.3709	0.2889	0.2318	0.4552	0.3695	0.2874	0.2319
0.5220	0.4272	0.3344	0.2699	0.5217	0.4269	0.3338	0.2683
0.5314	0.4355	0.3411	0.2755	0.5317	0.4345	0.3408	0.3008
0.5792	0.4772	0.3751	0.3043	0.5791	0.4771	0.3746	0.3033
0.5888	0.4857	0.3821	0.3102	0.5878	0.4848	0.3809	0.3105
0.6672	0.5557	0.4102	0.3343	0.6677	0.5552	0.4112	0.3347
0.7073	0.5922	0.4689	0.3852	0.7072	0.5912	0.4685	0.3869
0.7583	0.6391	0.5073	0.4189	0.7578	0.6388	0.5066	0.4203

capability of the trained Elman model is tested by 50 sets of unknown test data. A few samples of the test results are presented in Tables 3.6 and 3.7. These tables demonstrate the high degree of coherence with which the results obtained from the classical analysis tally with the results obtained from analysis of the practical networks using the depicted ANN-based model.

The training performance of the developed methodology is also shown in Figure 3.15 for ready reference.

TABLE 3.7

Comparison of Results Obtained by Classical Calculation and ANN-Based Model (Eastern Grid of India)

		Results Obtained from Classical Calculation			Results Obtained from ANN		
Bus No.	Reactive Demand (p.u.)	L Index	LQF	Voltage Mag (p.u.)	L Index	LQF	Voltage Mag (p.u.)
168	0.316	0.7386	0.6507	0.6019	0.7212	0.6581	0.6710
	0.458	0.6019	0.9071	0.5903	0.6597	0.9001	0.5870
	0.466	0.5903	0.9216	0.5743	0.5827	0.9197	0.5823
	0.476	0.5743	0.9399	0.5709	0.6091	0.9190	0.5890
	0.478	0.5709	0.9435	0.5673	0.6003	0.9510	0.5776
169	0.480	0.5673	0.9472	0.5618	0.5594	0.9312	0.5576
	0.483	0.5618	0.9527	0.5433	0.5498	0.9478	0.5399
	0.492	0.5433	0.9692	0.5363	0.5424	0.9701	0.5310
	0.495	0.5363	0.9747	0.5071	0.5915	0.9800	0.5071
	0.505	0.5071	0.9932	0.5034	0.5076	0.9898	0.5124
173	0.306	0.7034	0.6951	0.6994	06934	0.7201	0.7000
	0.375	0.6259	0.8038	0.7752	0.6088	0.7876	0.7812
	0.425	0.5679	0.8989	0.7124	0.5539	0.8880	0.7061
	0.509	0.4904	1.0008	0.4853	0.5185	1.0109	0.4989
	0.510	0.4853	1.0027	0.5885	0.4853	1.0987	0.5990

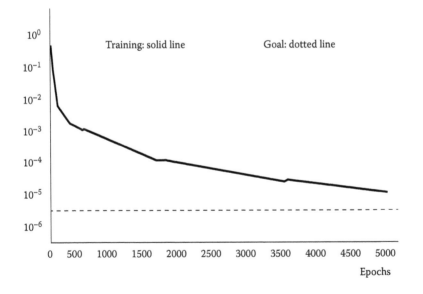

FIGURE 3.15

Training performance of developed ANN-based model.

It is also evident from these tables that the classical analysis and neural network analysis agree to a high degree with a few minor ignorable departures. However, the spectrum of advantages of the developed ANN-based model significantly scores over the negligible inaccuracy in its performance. The successful agreement of this aforesaid neural network methodology with classical results envisages the possibility of real-time prediction and correction of voltage stability.

3.5 Summary

A system enters a state of voltage instability when a disturbance causes a progressive and uncontrollable decline in voltage. Growing demand without matching expansion of generation and transmission facilities contributes to an increase of the frequency of these disturbances to continuously jeopardize the stability of the system. Hence, in the present-day power market scenario, it is quite imperative to maintain a constant surveillance on the system voltage profile and new methods are to be harnessed to control the same. The classical approaches of stability estimation require cumbersome and tedious off-line calculation for their inherent complexity. Thus, for fast and accurate estimation of control parameters, the classical approaches of stability determination are not suitable and hence cannot be adopted for on-line correction. Effective uses of ANN in this respect have been depicted in this chapter for real-time assessment of voltage instability. The simulation results of the adopted methodology are quite encouraging and optimistic.

Annotating Outline

- Voltage instability is one of the most frequent problems encountered by power networks.
- Investigation of voltage instability should be carried out on the weakest link of the network, so that the same methodology can be implemented in weaker or weak zones of the network. The weak buses have been identified employing a standard technique (Figures 3.12 and 3.13).
- The classical methods of voltage instability determination require cumbersome and tedious calculation involving a high degree of time complexity (Tables 3.3 and 3.4).

- Instead of determining the voltage stability directly with different indices as described in the classical method, a cost-effective artificially intelligent system has been developed to predict the same almost instantaneously with the training of the classical data set (Figure 3.14).
- The ANN-based model developed in this pursuit can be utilized for faster and accurate prediction of voltage instability to avoid its inevitable consequences (Tables 3.6 and 3.7). As an added benefit, this developed ANN-based model will not draw more investment cost, which may perturb the spot price of electricity in the power market.

4

Improvement of System Performances Using FACTS and HVDC

As the existing power system has not evolved at the rate of increase in demand and power transmission complexities, compensation in the form of advanced power electronic devices like flexible AC transmission system (FACTS) and high-voltage direct current (HVDC) is explored. Modern power systems suffer from poor stability margin, low efficiency and loadability limit. To exfoliate these abnormalities, by optimum utilization of existing FACTS devices and HVDC links, some flexible modeling and artificial neural network (ANN)-based training strategies have been developed to enable the independent system operator (ISO) to choose the most beneficial method of compensation. Different contingency sensitivity indices have also been personified to identify congestion and remove the same in pursuit of higher reliability and efficiency of the transmission network.

4.1 Introduction

Power system performance basically depends on stability margin, system efficiency and loadability. Though these factors are interrelated, individually they can predict the performance of a power network. A wider stability margin instigates formidable insulation against sudden and steady-state disturbances. It also increases the loadability of a network, which is an essential requirement of modern power systems. The stability margin can also be achieved by the improvement of the voltage profile of a system. On the other hand, the efficiency of a power system can be improved by reducing the power losses in the network. Hence, both voltage profile improvement and loss minimization are quite imperative to be implemented in power networks. Without disturbing the generation and load patterns, external devices can be incorporated in strategic locations to achieve better control over the power network. Flexible AC transmission system (FACTS) controllers narrow the gap between the uncontrolled and the controlled power system mode of operation, by providing additional degrees of freedom to control power flows and voltages at key locations of the network [288]. The primary limitations of the transmission of power are high active and reactive power loss on heavily loaded conditions and line outages. As proclaimed earlier, power networks can be modified to alleviate voltage instability or collapse by adding reactive power sources, i.e. shunt capacitors or FACTS devices. Unlike shunt capacitors, FACTS controllers can be connected as series or shunt compensators, depending upon the topology of the network. FACTS controllers such as STATCOM, SVC, TCSC, SSSC and UPFC are able to change the network parameters in a fast and effective way in order to achieve better system performance. Most of these FACTS devices, however, draw higher capacitive charging current, and hence limit power transfer capacity over long distances. There are also environmental concerns regarding the electrical and magnetic fields surrounding the overhead lines and AC cables. These limitations can be eliminated if direct current transmission is used. A DC transmission improves transmission capabilities, has lower losses and the transmission lengths are practically unlimited due to the elimination of capacitive currents. Moreover, during contingency or fault HVDC can insulate two interconnected systems from each other, thereby improving reliability. Though HVDC has numerous technical advantages over FACTS, its use is limited by the high investment cost associated with it. But, at present, deregulation tendency is driving conventional utilities to be more sensitive to new investment schemes. One of the main purposes of deregulation is to promote efficient use of cheap generation. However, owing to security issues, there are often constraints on the power transfer through transmission systems, causing overloading or congestion. It can also be caused by different levels of contingencies in interconnected networks. In this situation, the system will require more expensive generators, and therefore a power system is running in a sub-optimal mode in terms of power system economics. The optimal operating condition, however, can be achieved by changing the internal parameters of a network, such as transformer ratios and reactive power injections in buses, or by installing a FACTS or HVDC link, although these are quite obligatory for economic reasons.

This chapter establishes the applicability and effectiveness of these FACTS and HVDC links for the improvement of power network performance in terms of stability margin, line loss profile and loadability. The economic aspects of these external devices have also been optimized with the assistance of innovative algorithms and methodology to enhance system security considering all the possible working constraints.

4.2 Development of FACTS Controllers

Power flow control has traditionally relied on generator control, voltage regulation by means of tap changing and phase-shifting transformers and reactive power compensation switching. A phase-shifting transformer is used for the purpose of regulating active power in alternating current transmission networks. In practice, some of them are permanently operated with fixed angles, but in most cases their variable tapping facilities are actually made use of.

Series reactors are used to reduce power flow and short-circuit levels at designated locations of the network. Conversely, series capacitors are used to shorten the electrical length of the lines, hence increasing the power flow. In general, series compensation is switched on and off according to load and voltage conditions. For instance, in longitudinal power systems, series capacitive compensation is bypassed during minimum loading in order to avoid transmission line overvoltages due to excessive capacitive effects in the system. On the contrary, series capacitive compensation is fully utilized during maximum loading, aiming at increasing the transfer of power without subjecting transmission lines to overloads.

Until recently, these solutions served well the needs of the electricity supply industry. However, deregulation of the industry and difficulties in securing new rights of way have created the momentum for adapting new, radical technological developments based on high-voltage, high-current solid-state controllers.

Early developments of FACTS technology were in power electronic versions of the phase-shifting and tap-changing transformers. These controllers together with the electronic series compensator can be considered to belong to the first generation of FACTS equipment. The unified power flow controller, the static compensator and the interphase power controller are more recent developments. Their control capabilities and intended function are more sophisticated than those of the first wave of FACTS controllers. They may be considered to belong to a second generation of FACTS equipment. Shunt-connected thyristor-switched capacitors and thyristor-controlled reactors, as well as high-voltage direct current power converters, have been in existence for many years, although their operational characteristics resemble those of FACTS controllers.

FACTS controllers intended for steady-state operation are as follows [289]:

- Thyristor-controlled phase shifter (PS): This controller is an electronic phase-shifting transformer adjusted by thyristor switches to provide a rapidly varying phase angle.
- Load tap changer (LTC): This may be considered to be a FACTS controller if the tap changes are controlled by thyristor switches.
- Thyristor-controlled reactor (TCR): This is a shunt-connected, thyristor-controlled reactor, the effective reactance of which is varied in a continuous manner by partial conduction control of the thyristor valve.
- Thyristor-controlled series capacitor (TCSC): This controller consists of a series capacitor paralleled by a thyristor-controlled reactor in order to provide smooth variable series compensation.

- Interphase power controller (IPC): This is a series-connected controller comprising two parallel branches, one inductive and one capacitive, subjected to separate phase-shifted voltage magnitudes. Active power control is set by independent or coordinated adjustment of the two phase-shifting sources and two variable reactances. Reactive power control is independent of active power.
- Static compensator (STATCOM): This is a solid-state synchronous condenser connected in shunt with the AC system. The output current is adjusted to control either the nodal voltage magnitude or reactive power injected at the bus.
- Static synchronous series compensator (SSSC): This controller is similar to the STATCOM, but it is connected in series with the AC system. The output current is adjusted to control either the nodal voltage magnitude or the reactive power injected at one of the terminals of the series-connected transformer.
- Unified power flow controller (UPFC): This consists of a static synchronous series compensator (SSSC) and a STATCOM, connected in such a way that they share a common DC capacitor. The UPFC, by means of an angularly unconstrained series voltage injection, is able to control, concurrently or selectively, the transmission line impedance, the nodal voltage magnitude and the active and reactive power flow through it. It may also provide independently controllable shunt reactive compensation.

It is quite evident from the above discussion that the different topologies of FACTS devices are dedicated to different corresponding solutions of power network problems. One should be more selective in choosing the proper device for the eradication of particular limit violations of operational constraints. Hence, before incorporating a FACTS device, the network and its operational constraints should be thoroughly studied. Once the problem is identified, the proper topology can be selected and implemented. The selection and implementation process, however, is restricted by the high installation cost involved with these devices [290].

The application of FACTS controllers to the solution of steady-state operating problems is outlined in Table 4.1.

In order to assist power system engineers to assess the impact of FACTS equipment on transmission system performance, it has become necessary to write new power system software or to upgrade existing software. This has called for the development of a new generation of mathematical models for transmission systems and FACTS controllers, which had to be blended together, coded and extensively verified. This has been an area of intense research activity, which has given rise to a copious volume of publications.

From the operational point of view, FACTS technology is concerned with the ability to control, in an adaptive fashion, the path of the power flows throughout

TABLE 4.1

The Role of FACTS Controllers in Power System Operation

	Operating Problem	Corrective Action	FACTS Controllers
Voltage limits	Low voltage at heavy load	Supply reactive power	STATCOM, SVC
	High voltage at low load	Absorb reactive power	STATCOM, SVC, TCR
	High voltage following an outage	Absorb reactive power; prevent overload	STATCOM, SVC, TCR
	Low voltage following an outage	Supply reactive power; prevent overload	STATCOM, SVC
Thermal limits	Transmission circuit overload	Reduce overload	TCSC, SSSC, UPFC, IPC, PS
	Tripping of parallel circuits	Limit circuit loading	TCSC, SSSC, UPFC, IPC, PS
	Parallel line load sharing	Adjust series reactance	IPC, SSSC, UPFC, TCSC, PS
Loop flows	Post-fault power flow sharing	Rearrange network or use thermal limit action	IPC, SSSC, UPFC, TCSC, PS
	Power flow direction reversal	Adjust phase angle	IPC, SSSC, UPFC, PS

the network where before the advent of FACTS, high-speed control was very restricted. The ability to control the line impedance and nodal voltage magnitudes and phase angles at both the sending and receiving ends of the key transmission lines, with almost no delay, has significantly increased the transmission capabilities of the network while considerably enhancing the security of the system. In this context, computer-based power flow programs with FACTS controller modeling capability have been very useful tools for system planners and system operators to evaluate the technical and economical benefits of a wide range of alternative solutions offered by the FACTS technology.

In the following sub-sections the elementary series and shunt compensating FACTS devices have been modelled for their effective inclusion in power flow programs.

4.2.1 Modeling of Shunt Compensating Device

From an operational point of view, the static VAR compensator (SVC) behaves like a shunt-connected variable reactance, which either generates or absorbs reactive power in order to regulate the voltage magnitudes at the point of connection in the AC network. The firing angle control of the thyristor enables the SVC to have almost instantaneous speed of response to regulate voltage.

Conventional and advanced power flow models of SVCs are presented in this section. The advanced models depart from the conventional generator type representation of the SVC and are based instead on the variable shunt susceptance concept. In the latter case, the SVC state variables are combined with the nodal voltage magnitudes and angles of the network in a single frame of reference for unified, iterative solutions using Newton–Raphson method. Two models can be presented in this category [291]: the variable shunt susceptance model and the firing angle model.

4.2.1.1 Conventional Model of SVC

Early SVC models for power flow analysis treat the SVC as a generator behind an inductive reactance. The reactance accounts for the SVC voltage regulation characteristics.

A simpler representation assumes that the SVC slope is zero, an assumption that may be acceptable as long as the SVC operates within its design limits, but one that may lead to gross errors if the SVC is operating close to its limits. This point has been illustrated in Figure 4.1, with reference to the upper characteristic when the system is operating under low loading conditions [290]. If the slope is taken to be zero, then the generator will violate its minimum limit, point $AX_{SL=0}$. However, the generator will operate well within the limits, if the SVC voltage-current slope is taken into account at point A.

The SVC voltage-current slope can be represented by connecting the SVC model to an auxiliary bus coupled to the high-voltage bus by an inductive reactance consisting of the transformer reactance and the SVC slope, in per unit on the SVC base. The auxiliary bus is represented as a PV bus and the high-voltage bus is taken to be PQ. This model is shown schematically in Figure 4.2a. Alternatively, the SVC coupling transformer may be represented explicitly as shown in Figure 4.2b.

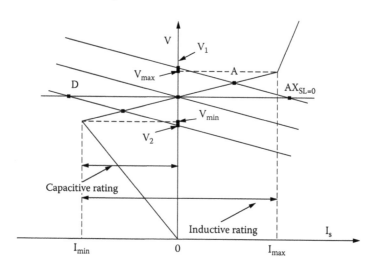

FIGURE 4.1
Static VAR compensator and power system voltage-current characteristics.

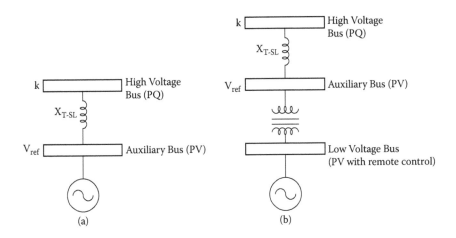

FIGURE 4.2
Conventional static VAR compensator power flow models: (a) slope representation and (b) slope and coupling transformer representation.

These SVC representations are quite straightforward but are invalid for operation outside the limits. In such cases, it becomes necessary to change the SVC representation to a fixed reactive susceptance, given by

$$B_{SVC} = \frac{Q_{\lim}}{V_{SVC}^2} \qquad (4.1)$$

where V_{SVC} is the newly fed voltage due to the reactive power limit $Q_{\lim}$ being exceeded.

The restriction applied by $Q_{\lim}$ can, however, be avoided by an advanced model discussed in the following part.

4.2.1.2 Shunt Variable Susceptance Model of SVC

In practice, the SVC can be seen as an adjustable reactance with either firing angle limits or reactance limits. The equivalent circuit shown in Figure 4.3 is used to derive the SVC non-linear power equations and the liberalized equations required by Newton's method.

With reference to Figure 4.3, the current drawn by the SVC is

$$I_{SVC} = jB_{SVC}V_k \qquad (4.2)$$

The reactive power drawn by the SVC, which is also the reactive power injected at bus k, is

$$Q_{SVC} = Q_k = -V_k^2 B_{SVC} \qquad (4.3)$$

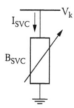

FIGURE 4.3
Variable shunt susceptance.

The linearized equation, where the equivalent susceptance B_{SVC} is taken to be a state variable, is as follows:

$$
\begin{bmatrix} \Delta P_k \\ \Delta Q_k \end{bmatrix}^{(i)} = \begin{bmatrix} 0 & 0 \\ 0 & Q_k \end{bmatrix}^{(i)} \begin{bmatrix} \Delta\theta_k \\ \Delta B_{SVC}/B_{SVC} \end{bmatrix}^{(i)} \tag{4.4}
$$

At the end of iteration (i), the variable shunt susceptance B_{SVC} is updated according to

$$
B_{SVC}^{(i)} = B_{SVC}^{(i-1)} + \left(\frac{\Delta B_{SVC}}{B_{SVC}}\right)^{(i)} B_{SVC}^{(i-1)} \tag{4.5}
$$

The changing susceptance represents the total SVC susceptance necessary to maintain the nodal voltage magnitude at the specified value.

Once the level of compensation has been computed, then the thyristor firing angle can be calculated. However, the additional calculation requires an iterative solution because the SVC susceptance and thyristor firing angle are non-linearly related.

4.2.1.3 Firing Angle Model of SVC

An alternative SVC model, which circumvents the additional iterative process, consists in handling the thyristor-controlled reactor (TCR) firing angle α as a state variable in the power flow formulation. The variable α has been designated here as α_{SVC} to distinguish it from the TCR firing angle α.

The positive sequence susceptance of the SVC can be written as follows:

$$
Q_k = \frac{-V_k^2}{X_C X_L}\left\{ X_L - \frac{X_C}{\pi}[2(\pi - \alpha_{SVC}) + \sin(2\alpha_{SVC})] \right\} \tag{4.6}
$$

From (4.6), the linearized SVC equation is given as

$$
\begin{bmatrix} \Delta P_k \\ \Delta Q_k \end{bmatrix}^{(i)} = \begin{bmatrix} 0 & 0 \\ 0 & \dfrac{2V_k^2}{\pi X_L}[\cos(2\alpha_{SVC}) - 1] \end{bmatrix}^{(i)} \begin{bmatrix} \Delta\theta_k \\ \Delta\alpha_{SVC} \end{bmatrix}^{(i)} \tag{4.7}
$$

At the end of iteration (i), the variable firing angle α_{SVC} is updated as follows:

$$\alpha_{SVC}^{(i)} = \alpha_{SVC}^{(i-1)} + \Delta\alpha_{SVC}^{(i-1)} \tag{4.8}$$

4.2.2 Modeling of Series Compensating Device

Two alternative power flow models to assess the impact of TCSC equipment in network-wide applications have been presented in this section. The simpler TCSC model exploits the concept of a variable series reactance. The series reactance is adjusted automatically, within limits, to justify a specified amount of active power flows through it. The more advanced model uses directly the TCSC reactance firing angle characteristic, given in the form of a non-linear equation. The TCSC firing angle is chosen to be the state variable in the Newton–Raphson power flow solution.

4.2.2.1 Variable Series Impedance Power Flow Model of TCSC

The TCSC power flow model presented in this section is based on the simple concept of a variable series reactance, the value of which is adjusted automatically to constrain the power flow across the branch to a specified value. The amount of reactance is determined efficiently using Newton's method. The changing reactance X_{TCSC}, shown in Figure 4.4, represents the equivalent reactance of all the series-connected modules making up the TCSC, when operating in either the inductive or the capacitive regions.

The transfer admittance matrix of the variable series compensator shown in Figure 4.4 is given by

$$\begin{bmatrix} I_k \\ I_m \end{bmatrix} = \begin{bmatrix} jB_{kk} & jB_{km} \\ jB_{mk} & jB_{mm} \end{bmatrix} \begin{bmatrix} V_k \\ V_m \end{bmatrix} \tag{4.9}$$

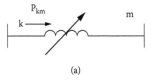

(a)

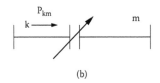

(b)

FIGURE 4.4
Thyristor-controlled series compensator equivalent circuit: (a) inductive and (b) capacitive operative regions.

For inductive operation

$$B_{kk} = B_{mm} = -\frac{1}{X_{TCSC}}; B_{km} = B_{mk} = \frac{1}{X_{TCSC}} \qquad (4.10)$$

And for capacitive operation the signs will be reversed.

The active and reactive power equations at bus k are

$$P_k = V_k V_m B_{km} \sin(\theta_k - \theta_m) \qquad (4.11)$$

$$Q_k = -V_k^2 B_{kk} - V_k V_m B_{km} \cos(\theta_k - \theta_m) \qquad (4.12)$$

For the power equations at bus m, the subscripts k and m are exchanged in (4.11) and (4.12).

In the Newton–Raphson solution these equations are linearized with respect to the series reactance. For the condition shown in Figure 4.4, where the series reactance regulates the amount of active power flowing from bus k to m at a value P_{km}^{reg}, the set of linearized power flow equations is

$$\begin{bmatrix} \Delta P_k \\ \Delta Q_k \\ \Delta P_m \\ \Delta Q_m \\ \Delta P_{km}^{X_{TCSC}} \end{bmatrix} = \begin{bmatrix} \frac{\partial P_k}{\partial \theta_k} & \frac{\partial P_k}{\partial |V_k|}|V_k| & \frac{\partial P_k}{\partial \theta_m} & \frac{\partial P_k}{\partial |V_m|}|V_m| & \frac{\partial P_k}{\partial X_{TCSC}}X_{TCSC} \\[1ex] \frac{\partial Q_k}{\partial \theta_k} & \frac{\partial Q_k}{\partial |V_k|}|V_k| & \frac{\partial Q_k}{\partial \theta_m} & \frac{\partial Q_k}{\partial |V_m|}|V_m| & \frac{\partial Q_k}{\partial X_{TCSC}}X_{TCSC} \\[1ex] \frac{\partial P_m}{\partial \theta_k} & \frac{\partial P_m}{\partial |V_k|}|V_k| & \frac{\partial P_m}{\partial \theta_m} & \frac{\partial P_m}{\partial |V_m|}|V_m| & \frac{\partial P_m}{\partial X_{TCSC}}X_{TCSC} \\[1ex] \frac{\partial Q_m}{\partial \theta_k} & \frac{\partial Q_m}{\partial |V_k|}|V_k| & \frac{\partial Q_m}{\partial \theta_m} & \frac{\partial Q_m}{\partial |V_m|}|V_m| & \frac{\partial Q_m}{\partial X_{TCSC}}X_{TCSC} \\[1ex] \frac{\partial P_{km}^{X_{TCSC}}}{\partial \theta_k} & \frac{\partial P_{km}^{X_{TCSC}}}{\partial |V_k|}|V_k| & \frac{\partial P_{km}^{X_{TCSC}}}{\partial \theta_m} & \frac{\partial P_{km}^{X_{TCSC}}}{\partial |V_m|}|V_m| & \frac{\partial P_{km}^{X_{TCSC}}}{\partial X_{TCSC}}X_{TCSC} \end{bmatrix} \begin{bmatrix} \Delta \theta_k \\ \frac{\Delta |V_k|}{|V_k|} \\ \Delta \theta_m \\ \frac{\Delta |V_m|}{|V_m|} \\ \frac{\Delta X_{TCSC}}{X_{TCSC}} \end{bmatrix}$$

$$(4.13)$$

where

$$\Delta P_{km}^{X_{TCSC}} = \Delta P_{km}^{reg} - \Delta P_{km}^{X_{TCSC}, cal} \qquad (4.14)$$

$\Delta P_{km}^{X_{TCSC}}$ is the active power mismatch for the series reactance, X_{TCSC}, given by

$$X_{TCSC} = X_{TCSC}^i - X_{TCSC}^{(i-1)} \qquad (4.15)$$

X_{TCSC} is the incremental change in series reactance, and $\Delta P_{km}^{X_{TCSC}, cal}$ is the calculated power.

The state variable X_{TCSC} of the series controller is updated at the end of each iterative step according to the following equation:

$$X_{TCSC}^{(i)} = X_{TCSC}^{(i-1)} + \left(\frac{\Delta X_{TCSC}}{X_{TCSC}} \right)^{(i)} X_{TCSC}^{(i-1)} \tag{4.16}$$

where the elements of the additional row and column of the modified Jacobian are derived as follows:

$$\frac{\partial P_{km}}{\partial \delta_k} = \frac{-|V_k V_m|}{X_{TCSC}} \cos(\delta_k - \delta_m) \tag{4.17}$$

$$\frac{\partial P_{km}}{\partial \delta_m} = \frac{|V_k V_m|}{X_{TCSC}} \cos(\delta_k - \delta_m) \tag{4.18}$$

$$\frac{\partial P_{km}}{\partial |V_k|} = \frac{-|V_m|}{X_{TCSC}} \sin(\delta_k - \delta_m) \tag{4.19}$$

$$\frac{\partial P_{km}}{\partial |V_m|} = \frac{-|V_k|}{X_{TCSC}} \sin(\delta_k - \delta_m) \tag{4.20}$$

$$\frac{\partial P_{km}}{\partial X_{TCSC}} = \frac{|V_k V_m|}{X_{TCSC}^2} \sin(\delta_k - \delta_m) \tag{4.21}$$

$$\frac{\partial P_k}{\partial X_{TCSC}} = \frac{-|V_k V_m|}{X_{TCSC}^2} \sin(\delta_k - \delta_m) \tag{4.22}$$

$$\frac{\partial P_m}{\partial X_{TCSC}} = \frac{|V_k V_m|}{X_{TCSC}^2} \sin(\delta_k - \delta_m) \tag{4.23}$$

$$\frac{\partial Q_k}{\partial X_{TCSC}} = \frac{-|V_k|^2}{X_{TCSC}^2} - \frac{|V_k V_m|}{X_{TCSC}^2} \cos(\delta_k - \delta_m) \tag{4.24}$$

$$\frac{\partial Q_m}{\partial X_{TCSC}} = \frac{-|V_m|^2}{X_{TCSC}^2} - \frac{|V_k V_m|}{X_{TCSC}^2} \cos(\delta_k - \delta_m) \tag{4.25}$$

4.2.2.2 Firing Angle Power Flow Model of TCSC

The model presented in Section 4.2.2.1 uses the concept of an equivalent series reactance to represent the TCSC. Once the value of the reactance is determined using Newton's method, then the associated firing angle can be calculated.

This makes engineering sense only in cases when all the modules making up the TCSC have identical design characteristics and are made to operate at equal firing angles. If this is the case, the computation of firing angle is carried out. However, such calculation involves an iterative solution since the TCSC reactance and the firing angle are non-linearly related. One way to avoid the additional iterative process is to use the alternative TCSC power flow model presented in this section [292].

The fundamental equivalent reactance $X_{TCSC(1)}$ as a function of firing angle of the TCSC module shown in Figure 4.5 is given as

$$X_{TCSC(1)} = -X_C + C_1\left[2(\pi - \alpha) + \sin(2(\pi - \alpha))\right]$$
$$- C_2 \cos^2(\pi - \alpha)\left[\varpi \tan(\varpi(\pi - \alpha)) - \tan(\pi - \alpha)\right] \tag{4.26}$$

where

$$X_{LC} = \frac{X_C X_L}{X_C - X_L}, C_1 = \frac{X_C + X_{LC}}{\pi}, C_2 = \frac{4X_{LC}^2}{X_L\pi} \text{ and } \varpi = \left(\frac{X_C}{X_L}\right)^{\frac{1}{2}} \tag{4.27}$$

The equivalent reactance $X_{TCSC(1)}$ in (4.26) replaces X_{TCSC} in (4.11) and (4.12), and the TCSC active and reactive power equations at bus k are

$$P_k = V_k V_m B_{km(f1)} \sin(\theta_k - \theta_m) \tag{4.28}$$

$$Q_k = -V_k^2 B_{kk(f1)} - V_k V_m B_{km(f1)} \cos(\theta_k - \theta_m) \tag{4.29}$$

where

$$B_{kk(1)} = -B_{km(1)} = B_{TCSC(1)} \tag{4.30}$$

For the case when the TCSC controls active power flowing from bus k to bus m, at a specified value, the set of linearized power flow equations is

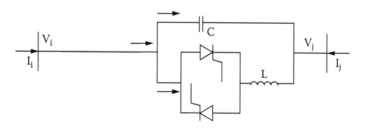

FIGURE 4.5
TCSC module.

$$
\begin{bmatrix}
\Delta P_k \\
\Delta Q_k \\
\Delta P_m \\
\Delta Q_m \\
\Delta P_{km}^{X_{TCSC}}
\end{bmatrix}
=
\begin{bmatrix}
\dfrac{\partial P_k}{\partial \theta_k} & \dfrac{\partial P_k}{\partial |V_k|}|V_k| & \dfrac{\partial P_k}{\partial \theta_m} & \dfrac{\partial P_k}{\partial |V_m|}|V_m| & \dfrac{\partial P_k}{\partial \alpha} \\[2ex]
\dfrac{\partial Q_k}{\partial \theta_k} & \dfrac{\partial Q_k}{\partial |V_k|}|V_k| & \dfrac{\partial Q_k}{\partial \theta_m} & \dfrac{\partial Q_k}{\partial |V_m|}|V_m| & \dfrac{\partial Q_k}{\partial \alpha} \\[2ex]
\dfrac{\partial P_m}{\partial \theta_k} & \dfrac{\partial P_m}{\partial |V_k|}|V_k| & \dfrac{\partial P_m}{\partial \theta_m} & \dfrac{\partial P_m}{\partial |V_m|}|V_m| & \dfrac{\partial P_m}{\partial \alpha} \\[2ex]
\dfrac{\partial Q_m}{\partial \theta_k} & \dfrac{\partial Q_m}{\partial |V_k|}|V_k| & \dfrac{\partial Q_m}{\partial \theta_m} & \dfrac{\partial Q_m}{\partial |V_m|}|V_m| & \dfrac{\partial Q_m}{\partial \alpha} \\[2ex]
\dfrac{\partial P_{km}^{X_{TCSC}}}{\partial \theta_k} & \dfrac{\partial P_{km}^{X_{TCSC}}}{\partial |V_k|}|V_k| & \dfrac{\partial P_{km}^{X_{TCSC}}}{\partial \theta_m} & \dfrac{\partial P_{km}^{X_{TCSC}}}{\partial |V_m|}|V_m| & \dfrac{\partial P_{km}^{X_{TCSC}}}{\partial \alpha_{TCSC}}
\end{bmatrix}
\begin{bmatrix}
\Delta \theta_k \\
\dfrac{\Delta |V_k|}{|V_k|} \\
\Delta \theta_m \\
\dfrac{\Delta |V_m|}{|V_m|} \\
\Delta \alpha_{TCSC}
\end{bmatrix}
$$

$$\tag{4.31}$$

where $\Delta P_{km}^{X_{TCSC}}$ is given by

$$\Delta P_{km}^{X_{TCSC}} = \Delta P_{km}^{reg} - \Delta P_{km}^{X_{TCSC},cal} \tag{4.32}$$

$\Delta P_{km}^{X_{TCSC}}$ is the active power mismatch for the TCSC module. $\Delta \alpha_{TCSC}$ is given by

$$\Delta \alpha_{TCSC} = \alpha_{TCSC}^{(i+1)} - \alpha_{TCSC}^{(i)} \tag{4.33}$$

$\Delta \alpha_{TCSC}$ is the incremental change in TCSC firing angle at the ith iteration, and $\Delta P_{km}^{X_{TCSC},cal}$ is the calculated power.

4.3 Prologue of High-Voltage Direct Current (HVDC) System

Application of electricity originally started with the use of direct current. The first central electric station was installed by Edison in New York in 1882 supplying power at 110 V DC. The invention of the transformer and induction motor and the concept of three-phase AC around 1890 initiated the use of AC. The advantages of three-phase AC almost eliminated the use of DC systems expect for some special applications in electrolytic processes and adjustable-speed motor drives.

Today DC transmission has staged a comeback in the form of HVDC transmission to supplement the high-voltage alternating current (HVAC) transmission system. The first commercially used HVDC link (20 MW, 100 kV) in the world was built in 1954 between the mainland of Sweden and the island of Gotland [293]. Since then, the technique of power transmission by HVDC has been continuously developing. DC transmission is an effective means to improve system performance. It is mainly used to complement AC systems rather than to displace them. In India, the first HVDC 810 km long distance overhead line was from Rihand to Delhi (500 kV, 1500 MW) for bulk power transmission. At present, the world has over 60 HVDC schemes in operation for a total capacity of more than 66,000 MW, and the growth of DC transmission capacity has reached an average of 2500 MW/year [293].

Incorporation of HVDC in an existing power system network can enhance the performance of the same in the following ways:

1. Interconnection of the systems of the same frequency through a zero-length DC link (back-to-back connection). This does not require any DC transmission line and the AC lines can be terminated on the rectifier and inverter, which are connected back to back (Figure 4.6). This helps in interconnecting two AC systems without increasing their fault levels.

2. HVDC links are used to evacuate power from remote power stations to the load centres situated several hundred kilometres away. If there are faults in the AC network, this will not trip the units at the power station since the synchronous DC link insulates the power station from the AC system.

3. Stabilizing the AC system by modulating DC power flow.

4. Interconnection between power systems or pools. For smooth interchange of the power between neighbouring grids, irrespective of voltage and frequency fluctuations, such a link ensures the retention of the tie under the most stringent conditions of the constituent grids.

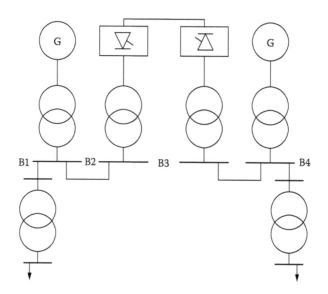

FIGURE 4.6
Back-to-back HVDC connection.

5. High-power underground distribution system feeders. It is found that DC may be cheaper at distances greater than approximately 50 km with a power level of 1000 to 2000 MW. For this large amount of power transfer with AC, it needs forced cooling due to the higher amount of heat produced. Also, there are increased dielectric losses at extra-high voltage (EHV) AC.

6. HVDC links do not increase the short-circuit level of the connected systems. Faults and oscillations do not transfer across HVDC interconnected systems. They act as firewalls against cascading outages.

7. Multi-terminal configurations are simple to engineer. They are ideal for offshore grids interconnecting wind farms.

8. Power reversal is also possible by adjusting the DC voltage at both converter stations.

As stated above, the offerings of HVDC are quite considerable. Hence, power markets worldwide are emerging with HVDC solutions to power network problems. Figure 4.7 summarizes these technologies.

However, the cost of terminal equipment is much more in the case of DC (converting stations) than in the case of AC (transformer/substation). Figure 4.8 shows the variation of cost of power as a function of transmission distance. The point of intersection P is called a breakeven point, which shows that a transmission distance of more than 650 km is only preferable to use with DC; otherwise, AC should be used.

But, considering the uncountable advantages, research and development work is under way to provide a better understanding of the performance of HVDC links to achieve more efficient and economic designs of the thyristor valves and related equipment and to justify the use of alternative AC/DC system configurations.

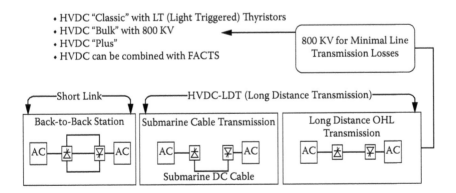

FIGURE 4.7
Different emerging technologies of HVDC.

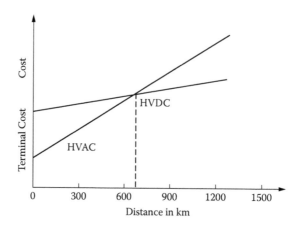

FIGURE 4.8
Variation of cost of power as a function of transmission distance.

In the subsequent sub-section, modeling of the DC link is explained in considerable detail, followed by application of the HVDC link in an AC network to enhance the system performance.

4.3.1 Modeling of DC Link

Considerable research works are under way to provide a better performance of HVDC links to achieve more efficiency of related equipment and to justify the use of the AC/DC system configuration. Figure 4.9 shows the schematic diagram of the basic model of the DC link interconnecting the bus bars, where V is the converter terminal bus bar nodal voltage, E is voltage at the converter transformer secondary, I is the current at the converter transformer secondary, γ and δ are the firing and extinction angles of converters, n is the transformer tap ratio and V_d, I_d are the direct voltage and current at the link.

The DC link model equations relating the variables are as follows [294]:

$$V_{dr} = V_{di} + R_{dc}I_d \tag{4.34}$$

$$P_{dr} = V_{dr}I_d \tag{4.35}$$

$$P_{di} = V_{di}I_d \tag{4.36}$$

$$V_{dr} = kn_rV_r\cos\gamma - \frac{3}{\pi}X_rI_d \tag{4.37}$$

$$V_{di} = kn_iV_i\cos\delta - \frac{3}{\pi}X_iI_d \tag{4.38}$$

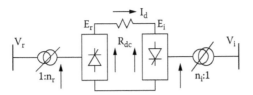

FIGURE 4.9
Schematic diagram of DC link.

Using the specified rectifier end power or inverter end power, inverter bus voltage, minimum values of firing and extinction angles, all the DC link variables can be obtained by using (4.34) to (4.38). With all DC variables, the real and reactive powers can be calculated by the following equations:

$$S(L) = V_{d(L)}I_d + jV_{d(L)}I_d \tan\theta(L) \tag{4.39}$$

$$\tan\theta(L) = \frac{\text{sgn}(L)2U + \sin 2\beta(L) - \sin(\beta(L)+U)}{\cos 2\beta(L) - \cos 2(\beta(L)+U)} \tag{4.40}$$

$$\{\text{If } L \text{ is rectifier, } \beta(L) = \gamma \text{ and } \text{sgn}(L) = 1; \text{ If } L \text{ is inverter, } \beta(L) = \delta \text{ and } \text{sgn}(L) = -1\}$$

where

$$U = \cos^{-1}\left[\frac{2V_{d(L)}}{V_0(L)} - \cos\beta(L)\right] - \beta(L) \tag{4.41}$$

$$V_0(L) = \frac{3\sqrt{2}}{\pi}V(L) \tag{4.42}$$

where $V(L)$ is the alternating voltage at the converter terminals.

The model developed above can alter the magnitude as well as direction of power flow only by changing the value of the firing angle. Hence, it is quite imperative to test the applicability of this model in a practical system. The following section focuses on incorporation of this model into power networks.

4.4 Improvement of System Performance Using FACTS and HVDC

In this section, the improvement in performance of the weak link of the system has been illustrated using a (1) shunt compensator, (2) series compensator and (3) parallel HVDC link. The IEEE 14 and IEEE 30 bus test systems and a practical 203 bus-265 line power network have been adopted for these methodical studies.

4.4.1 Improvement of Voltage Profile of Weak Bus Using SVC

Effective utilization of SVC in restoring the system voltage profile depends upon the installation of the same at an appropriate location.

The SVC connected in this fashion should be self-regularized and should adapt to the inevitable changes taking place in the network. Being a FACTS device, its sensitivity should be very high and the firing angle of the device should change accordingly in case of any limit violation. The flowchart presented in Figure 4.10 elaborates the implementation of a self-regulated SVC in stressed conditions of loading. As shown, the algorithm monitors the minimum value of the bus voltage and, by altering only the triggering angle of solid-state switches, convalesces the same.

Table 4.2 depicts the test results of this algorithm with various loading stresses applied on the weakest bus (by using (3.41)). The work models the

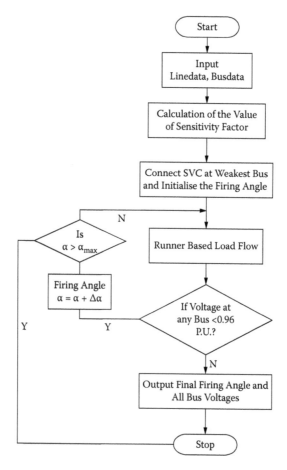

FIGURE 4.10
Flowchart for solution methodology.

TABLE 4.2

Effectiveness of SVC in Restoring Bus Voltage Profile under Excessive Loading Stress for IEEE 30 Bus System

Bus No.	Reactive Demand (p.u.)	Minimum Bus Voltage without SVC (p.u.)	Minimum Bus Voltage with SVC (p.u.)	Firing Angle of SVC (radian)
30	0.019	0.95357	0.96214	2.1031
	0.033	0.94350	0.96056	2.1190
	0.047	0.93310	0.96165	2.1207
	0.061	0.92248	0.96571	2.1296
	0.075	0.91158	0.97298	2.1386
	0.089	0.90032	0.96197	2.1387
	0.103	0.88881	0.97265	2.1478
	0.117	0.87689	0.96117	2.1479
	0.131	0.86456	0.97569	2.1573
	0.145	0.85179	0.96369	2.1574
	0.159	0.83852	0.98282	2.1671
	0.173	0.82468	0.97032	2.1673
	0.187	0.81021	0.99510	2.1775
	0.201	0.79501	0.98220	2.1778
	0.215	0.77896	0.96891	2.1781
	0.229	0.76189	0.99351	2.1893
	0.243	0.74360	0.98806	2.1900

SVC in such a way that only by trivial alteration of firing angle (within the operating limit) can the voltage profile of the buses be recovered. This recovery is essential as by maintaining the voltage profile, the maximum power transfer capabilities of the lines can be sustained and efficiency of the transmission can be increased with rated operational conditions.

The study remains incomplete without implementing this algorithm in practical systems. Henceforth, a 203 bus-265 line power network of the eastern grid of India has been adopted for experimentation. As this network is extended to a large area, instead of considering a single weak bus, three different case studies have been carried out for three different locations of loading stress. The position of SVC, considering its economic impact, however, remains the same for all three cases. In Table 4.3, the test results have been summarized.

Contingencies are inadvertent conditions that may appear in the system to create undesired limit violations. The applicability of the work has been extended up to maintaining system equilibrium even in the case of contingencies. Figures 4.11 and 4.12 show the contingency sensitivity factor (CSF) (developed in (4.43)) for different lines in healthy conditions of the system. It is quite evident that a higher value of this factor will correspond to the vulnerability of the lines in terms of contingency, and exclusion of most

TABLE 4.3

Effectiveness of SVC in Restoring Bus Voltage Profile under Excessive Loading Stress for Eastern Grid of India

Bus No.	Reactive Demand (p.u.)	Minimum Bus Voltage without SVC (p.u.)	Minimum Bus Voltage with SVC (p.u.)	Firing Angle of SVC (radian)
168	0.200	0.90991	0.94000	2.1508
	0.230	0.90988	0.94001	2.1653
	0.245	0.90988	0.94862	2.1800
	0.260	0.90984	0.94003	2.1801
	0.290	0.90980	0.94006	2.1954
	0.320	0.90976	0.94002	2.1957
	0.335	0.90974	0.94008	2.2117
	0.350	0.90971	0.94006	2.2120
	0.380	0.90966	0.94002	2.2128
	0.410	0.90961	0.94008	2.2311
	0.425	0.90958	0.94006	2.2322
	0.440	0.90955	0.94004	2.2337
	0.500	0.90938	0.94001	2.2514
169	0.183	0.90091	0.94000	2.1508
	0.200	0.90088	0.94002	2.1652
	0.215	0.90086	0.94107	2.3174
	0.230	0.89983	0.94297	2.4302
	0.245	0.89981	0.94630	2.5451
	0.260	0.89978	0.95383	2.6785
	0.275	0.89976	0.95380	2.6785
	0.290	0.89973	0.95378	2.6785
	0.305	0.89970	0.95376	2.6786
173	0.150	0.90091	0.94000	2.1508
	0.165	0.90090	0.94003	2.1652
	0.200	0.90086	0.94005	2.1799
	0.220	0.89984	0.94297	2.4302
	0.230	0.89983	0.94413	2,4765
	0.240	0.89982	0.94632	2.5452
	0.260	0.89980	0.95385	2.6786
	0.275	0.89978	0.95383	2.6786
	0.300	0.89975	0.95381	2.6787
	0.320	0.89973	0.95379	2.6787
	0.325	0.89972	0.95379	2.6787

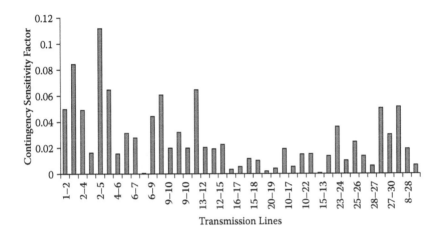

FIGURE 4.11
CSFs for different transmission lines of IEEE 30 bus system.

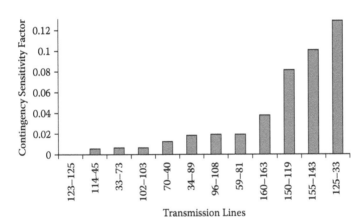

FIGURE 4.12
CSFs for different transmission lines of the eastern grid of India.

vulnerable lines of the system will represent the worst possible contingencies in terms of line flow.

$$CSF = \frac{P_{ij}}{P_{ij(max)}} \tag{4.43}$$

where P_{ij} and $P_{ij(max)}$ are the power flow and maximum power flow between the ith and j^{th} buses, respectively.

Accordingly, the contingencies are affected and the self-regulating capability of the algorithm is quite observable from Table 4.4. SVC, in the case of even multiple contingencies, can allow larger clearing time by maintaining the

TABLE 4.4

Restoration of Bus Voltage during Contingency in IEEE 30 Bus System

Lines Tripped	Minimum Bus Voltage without SVC (p.u.)	Minimum Bus Voltage with SVC (p.u.)	Firing Angle of SVC (radian)
1–2; 4–6	0.71688	0.96550	2.1909
1–2; 4–6; 4–12	0.73952	0.98745	2.1903
9–10; 15–18; 22–24	0.68904	0.96804	2.1932
1–2; 4–6; 4–12; 10–21	0.72837	0.97988	2.1907
4–6; 4–12; 10–21; 22–24	0.72916	0.98105	2.1907
1–2; 4–6; 4–12; 10–21; 16–17	0.71916	0.97105	2.1942
1–2; 4–6; 4–12; 10–21; 22–24	0.69567	0.97423	2.1929

TABLE 4.5

Restoration of Bus Voltage during Contingency in Eastern Grid of India

Lines Tripped	Minimum Bus Voltage without SVC (p.u.)	Minimum Bus Voltage with SVC (p.u.)	Firing Angle of SVC (radian)
34–89	0.66286	0.95247	2.2272
59–81	0.66098	0.95226	2.2277
123–125	0.66449	0.95283	2.2268
160–163	0.66331	0.95253	2.2271
150–119	0.65886	0.95313	2.2286
155–143	0.66104	0.95212	2.2277
125–33	0.66309	0.95252	2.2272
114–45	0.66340	0.95255	2.2271
96–108	0.66337	0.95254	2.2271
102–103	0.66346	0.95256	2.2271
102–103; 33–73	0.66336	0.95254	2.2275
34–89; 70–40	0.66344	0.95261	2.2273
34–89; 102–103	0.66278	0.95245	2.2272
160–163; 145–115	0.67218	0.95217	2.2301
160–163; 121–137	0.66105	0.95222	2.2277

system voltage without altering the generation pattern or subjecting the loads to curtailment. The use of this algorithm with the model-developed SVC can ensure not only reliability but also quality of the voltage profile in an inadvertent state of the system. The practical system of 203 buses has also been tested to give promising results to deploy the efficacy of the work. Table 4.5 stands as an alibi for implementation of this algorithm in a practical system.

The algorithm can produce the firing angle corresponding to each of the feasible loading stresses, but each time it has to check the results by running load flow. For faster implementation and updating of firing angle an artificially intelligent model can be trained to produce the optimum value of the firing angle for any loading stress without running the load flow. This not only saves valuable time for planning but also makes the system fast responding with vulnerability. The severity of any contingency can be avoided by real-time corrective action. With the faster prediction of firing angle, the system will be able to heal itself by sustaining its operating point.

4.4.2 Application of ANN for the Improvement Voltage Profile Using SVC

This section presents an ANN-based technique for upgradation of the voltage profile in the weak bus of a multi-bus power network with the assistance of SVC. In this methodology, a neural method (Elman network)-based algorithm has been developed (1) to determine the degree of weakness of the weak load bus with the change in load magnitude and (2) to find out the optimum firing angles of the SVC in order to improve the voltage status of the weak load bus. The developed ANN-based technique has been applied on the IEEE 30 bus system and is found to be in excellent agreement with the result obtained from the classical method.

The ANN-based model, as stated, has two aspects:

> Application I: For different load values, the ANN algorithm has attempted to predict the voltage magnitude, corresponding angle and $\frac{\partial Q}{\partial V}$ values for the second weakest bus.
>
> Application II: For different load values, the developed ANN model is able to find out the optimum firing angle values at which SVC should be fired to maintain a specified bus voltage of that particular bus.

Formulation of the problem and the proposed solution methodology for obtaining the desired solution has been depicted in the flowchart in Figure 4.13.

Two different network topologies have been used for the above-mentioned two applications. The network topology and parameters have been optimized by iterative checking to obtain best results. The network details are specified in Table 4.6.

The model has been trained with 100 (for application I) and 120 (for application II) sets of input training vectors to meet convergence criteria for the two applications that are achieved in 3500 and 5000 iterations, respectively, with a training goal of 0.0001. After attaining the training goal, the network correlates between the changing load pattern and the corresponding values of sensitivity index, voltage magnitudes and angles and optimum firing angles of SVC. A few samples of the test results are presented in Tables 4.7 and 4.8. The agreement between computed and estimated results is good (Figure 4.14).

The ability of this ANN-based model to predict the most feasible firing angle is thus incontestable. This model not only saves time but also makes the SVC-based voltage profile improvement system more reliable.

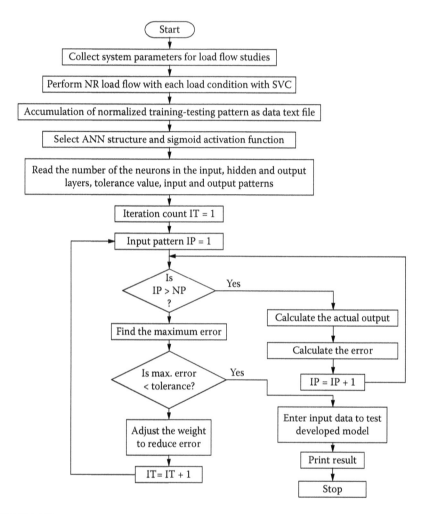

FIGURE 4.13
Developed ANN-based solution methodology.

TABLE 4.6

Design Parameters of Developed Neural Network

	Application I	Application II
Training vector	100	120
Testing vector	50	60
Input neurons	2	3
Hidden layers	1	1
Output neurons	3	3
Neurons in hidden layer	100	100

TABLE 4.7

Comparison of Results Obtained by Conventional Computation and ANN Model (for Application I)

Reactive Demand (p.u.)	Results Obtained from Classical Computation			Results Obtained from ANN Model		
	$\delta Q/\delta V$	V_{mag} (p.u.)	V_{angle} (radian)	$\delta Q/\delta V$ (p.u.)	V_{mag} (p.u.)	V_{angle} (radian)
0.019	2.9003	0.9681	0.2178	2.9046	0.9689	0.1982
0.047	2.8526	0.9572	0.2152	2.8489	0.9546	0.2222
0.061	2.8108	0.9476	0.212	2.8094	0.9407	0.2577
0.075	2.7969	0.9444	0.2122	2.7955	0.9443	0.1962
0.089	2.7824	0.9512	0.2114	2.7773	0.9527	0.2241
0.103	2.7680	0.9379	0.2106	2.7673	0.9384	0.2099
0.117	2.7535	0.9346	0.2099	2.7587	0.9382	0.2057
0.131	2.7462	0.9329	0.2095	2.7460	0.9275	0.2106
0.165	2.7316	0.9295	0.2087	2.7307	0.9269	0.2096
0.209	2.7019	0.9228	0.2071	2.7043	0.9271	0.1999
0.273	2.6717	0.9158	0.2056	2.6727	0.9191	0.2011
0.275	2.6487	0.9106	0.2044	2.6488	0.9124	0.2027
0.280	2.6332	0.971	0.2036	2.6381	0.9075	0.2037
0.290	2.5938	0.8980	0.2016	2.5932	0.8966	0.1986

TABLE 4.8

Comparison of Results Obtained by Conventional Computation and ANN Model (for Application II)

Reactive Demand (p.u.)	Results Obtained from Classical Computation			Results Obtained from ANN Model		
	V_{mag} (p.u.)	V_{angle} (radian)	SVC Firing Angle (degree)	V_{mag} (p.u.)	V_{angle} (radian)	SVC Firing Angle (degree)
0.019	0.9810	0.2210	122.0000	1.0072	0.1923	121.9279
0.033	0.9824	0.2213	122.5000	1.0904	0.1611	122.4701
0.047	0.9867	0.2224	123.5000	1.0166	0.1743	123.4492
0.061	0.9866	0.2224	124.3220	0.9947	0.2156	124.2212
0.075	0.9868	0.2224	124.6000	0.9862	0.2225	124.5828
0.089	0.9864	0.2223	124.8000	0.9788	0.2301	124.7855
0.103	0.9860	0.2222	125.0000	0.9727	0.2304	125.0493
0.117	0.9861	0.2222	125.3000	0.9675	0.2200	125.3099
0.131	0.9862	0.2222	125.6000	0.9632	0.2239	125.5918
0.165	0.9863	0.2223	125.9000	0.9596	0.2272	125.8985
0.209	0.9862	0.2222	131.0000	0.9991	0.2196	130.9946
0.273	0.9864	0.2223	137.4000	0.9859	0.2233	137.3637

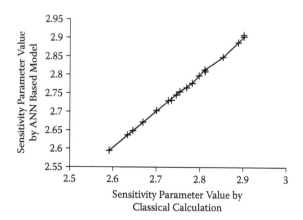

FIGURE 4.14

Test of ability of developed neural network-based model.

This methodology, however, may require large data storage space, which may not be availed for planning studies. Hence, the following work focuses on self-healing technologies like TCSC and HVDC and goes on to develop some algorithms that can make the transmission system more flexible to cope with stresses and contingencies without the help of any trained model.

4.4.3 Application of TCSC and HVDC for Upgrading of Cost-Constrained System Performance

This sub-section presents a novel approach to improve the voltage profile and loss minimization of a power network under stressed and contingent conditions [307]. In doing so, the weakest line (using (4.43)) in the system has been found and then that line has been supported by a TCSC in series and a parallel HVDC link separately. Comparison of simulated results indicates that HVDC is the best option for performance improvement for the weakest link, though the investment cost of the same is very high.

In this section, to study the performance of TCSC and the HVDC link separately with the weakest line of the power network, the cases below have been followed:

Determination of the weakest link in the system under stressed and contingent conditions

Performance of TCSC and the HVDC interconnection link separately in stressed conditions

Performance of TCSC and the HVDC interconnection link during contingent conditions

Cost comparison of TCSC and the HVDC link

4.4.3.1 Determination of the Weakest Link in the System under Stressed and Contingent Conditions

At the outset, the weakest link of the IEEE 14 bus test system has been determined by CSF. During contingency this interconnection may be that particular link that requires reactive power support the most. Since it is not an economically viable option to connect several interconnecting links with TCSC/HVDC support, the most overloaded link needs to be chosen to account for reactive power sensitivity during contingencies. To determine the most overloaded line during contingencies, several cases have been studied and CSFs for the four worst cases of contingencies have been calculated, as shown in Table 4.9.

Test cases suggest that the line interconnecting buses 13 and 14 is the most overloaded link, and hence TCSC/HVDC support should be placed in parallel with this existing weakest link. During contingency or stressed conditions this parallel line with TCSC/FACTS support may be turned on to relieve the lines from overloading and improve the operating conditions of interconnected systems. Before considering the investment cost associated with TCSC and HVDC, one must consider the overload or congestion relief of the lines with respect to secure operation of the network in contingent conditions. The long-term effects of the incorporation of these links can prove to be efficient in controlling locational marginal prices and congestion management cost.

TABLE 4.9

Contingency Sensitivity Factors (CSFs) of Different Lines for IEEE 14 Bus System

Transmission Lines	Tripped Line 1–5 (CSF)	Tripped Line 1–2 (CSF)	Tripped Line 9–14 (CSF)	Tripped Line 11–12 (CSF)
1–2	0.531	1.000	0.002	0.001
1–5	1.000	2.442	0.009	0.001
2–3	0.185	0.342	0.006	0.001
2–4	0.486	0.964	0.013	0.004
2–5	0.882	1.763	0.023	0.004
4–5	0.592	1.188	0.117	0.027
4–7	0.034	0.099	0.189	0.044
4–9	0.033	0.102	0.187	0.044
5–6	0.035	0.105	0.206	0.045
6–11	0.072	0.180	0.409	0.085
6–12	0.020	0.081	0.400	0.075
6–13	0.043	0.146	0.653	0.146
7–9	0.034	0.099	0.189	0.044
9–10	0.235	0.578	1.327	0.275
9–14	0.067	0.227	1.000	0.227
10–11	0.114	0.279	0.641	0.133
12–13	0.149	0.601	2.953	0.564
13–14	0.302	1.008	4.559	1.000

4.4.3.2 Performance of TCSC and the HVDC Interconnection Link Separately in Stressed Conditions

To study the performance of TCSC and the HVDC link separately with the weakest line of the power network, the cases below have been followed: performance of the weakest line without TCSC and the HVDC link, performance of the weakest line with TCSC in series and performance of the weakest line with HVDC links.

The TCSC has been connected with the weak link, and its impact on steady-state voltage stability has been observed with much higher values of active and reactive loads at bus numbers 13 and 14. On the other hand, a HVDC link with the characteristics shown in Table 4.10 has been connected in parallel with the same weak line to illustrate the improvements of operating conditions with an AC-DC link. To incorporate the AC-DC link in the system, the sequential approach for integration of the HVDC link in an AC load flow program has been followed. The voltage profiles for the cases are shown in Figure 4.15.

TABLE 4.10

DC Link Characteristics

	Rectifier	Inverter
Bus number	13	14
Commutation reactance (p.u.)	0.126	0.7275
Minimum control angle (degree)	7	10
Filter admittance (p.u.)	0.4902	0.6301
Resistance of DC line (p.u.)	0.00334	
DC power flow setting (p.u.)	0.4857	
DC voltage (p.u.)	1.284	

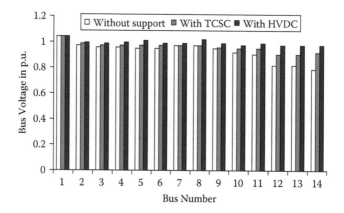

FIGURE 4.15
Voltage profile with and without compensation under stressed conditions.

Table 4.11 shows the DC link result. From the voltage profile of the system, improvement with TCSC and the HVDC interconnecting link has been observed. This is an important aspect of interconnection via HVDC or the TCSC link in deregulated power systems where a continuous increase of demand is expected and the links should be flexible enough to cope with this variation.

Table 4.12 depicts the values of the voltage stability indicator (VSI) for all three cases and also validates the above fact. It is worth mentioning that when several systems are interconnected, as in the case of deregulation, the overall system stability is bound to get vulnerable owing to a small disturbance in a part of the system. Therefore, improvement of voltage stability is highly desired in power systems with deregulation. As shown with the HVDC link, the system is least vulnerable to disturbances.

Modern electricity markets are expected to be governed by spot price of electricity. This spot price can be optimized by reducing power loss (both active and reactive), and hence production cost of units generated.

Table 4.13 represents the comparison of active and reactive power losses with and without the inclusion of TCSC and the HVDC link separately.

TABLE 4.11

Result of DC Link

	Rectifier	Inverter
Value of DC voltage	1.28448	1.28400
Value of transformer tap position	1.01150	1.03330
Value of control angles (degree)	16.22617	21.19778
Value of real power flow (p.u.)	0.18577	0.18570
Value of reactive power flow (p.u.)	3.38950	−0.83118
Value of power factor	0.99991	0.98609
Value of DC current (p.u.)	0.14463	

TABLE 4.12

Comparison of VSI Values under Loading Stress

Stability Indicator	Without FACT/HVDC	With TCSC	With HVDC
VSI	0.1323	0.0705	0.0340

TABLE 4.13

Comparison of Losses under Loading Stress

Without FACTS/HVDC		With TCSC		With HVDC	
Active Loss (p.u.)	Reactive Loss (p.u.)	Active Loss (p.u.)	Reactive Loss (p.u.)	Active Loss (p.u.)	Reactive Loss (p.u.)
0.278	0.5537	0.2482	0.4866	0.2133	0.4385

Though TCSC can cause a considerable reduction in total loss, HVDC can perform even better than TCSC. During the consideration of high investment cost of HVDC and TCSC, long-term benefits of reduced power loss should also be considered.

4.4.3.3 Performance of TCSC and the HVDC Interconnection Link during Line Contingency

This comparative analysis remains incomplete without studying the behavioural improvements of the system with the HVDC link or TCSC in case of disturbances like line contingency. With the 12-13 line trip, load flow has been carried out with and without TCSC and HVDC support. A comparison of voltage profile, stability and power loss with this line contingency is depicted in Figure 4.16 and Tables 4.14 and 4.15, respectively. It can be

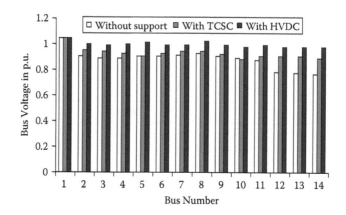

FIGURE 4.16
Voltage profile with and without compensation during contingency.

TABLE 4.14

Comparison of VSI Values during Contingency

Stability Indicator	Without FACT/HVDC	With TCSC	With HVDC
VSI	0.1739	0.07401	0.03580

TABLE 4.15

Comparison of Losses during Contingency

Without FACTS/HVDC		With FACTS		With HVDC	
Active Loss (p.u.)	Reactive Loss (p.u.)	Active Loss (p.u.)	Reactive Loss (p.u.)	Active Loss (p.u.)	Reactive Loss (p.u.)
0.2915	0.5872	0.2635	0.5247	0.2149	0.4400

observed that with TCSC or the HVDC link the voltage profile is much better than without their inclusion. The HVDC-AC link offers the best stability margin compared to the other two cases. It can be inferred that even in the case of line outage, the stability of the system remains the least affected with an integrated AC-DC link. Hence, system interconnection with HVDC or TCSC can raise the reliability of the system, which is an important issue of any power system regardless of deregulation. Moreover, with a better voltage profile and stability margin, the loading margins can also be increased.

4.4.3.4 Cost Comparison of TCSC and the HVDC Link

From the above discussions, it can be established that the integrated HVDC-AC system for the weakest link is the best option to achieve better system performance, although the investment cost of the HVDC link is sufficiently high. Figure 4.17 shows the cost comparison of these devices as a percentage of production cost without their support. In deregulated environments, customers are willing to pay more for improved power system security, reliability and better operating conditions. Hence, use of the HVDC link in parallel with the weakest AC link for improvement of system performance under contingent and stressed conditions can prove its cost-effectiveness in the long run.

The applicability and effectiveness of TCSC and the HVDC link in power system networks have thus been depicted in this work. All test results indicate that the performance of the weakest link has been improved with a parallel HVDC link. The link may also provide a certain degree of overload relief, which is quite essential for congestion management in deregulated environments of power systems. A cost comparison has also been provided with the IEEE 14 bus test system, which shows that investment cost of the HVDC link is higher than FACTS devices. But in deregulated environments, power system security, reliability and meeting the consumer target are more

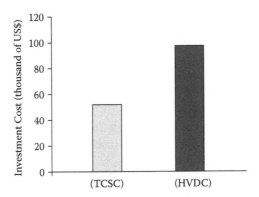

FIGURE 4.17
Cost comparisons of FACTS and HVDC link.

important issues to consider. Therefore, the efforts made in these works are expected to create a strategic guideline for future deregulation of power systems with respect to utilizing this HVDC link for interconnection.

4.5 Summary

Normally an electric power system should be operated within its operating constraints such as voltage limit, thermal limit, line loss limit, angle and voltage instability limits. Uncontrollable power flows may result in violations of these constraints. For effective utilization of the power network, new power system-controlled technologies such as FACTS devices and HVDC are quite imperative to be incorporated in the present-day power network. As these technologies demand high investment cost, their optimized employment is highly desired in power industries. In this respect, the endeavour of this chapter has been to develop unique modeling techniques and algorithms for cost-effective application of the available FACTS and HVDC technologies. In this pursuit, the research works have also deployed ANN for fast and more accurate determination of controlled parameters. As the future power market is expected to heavily rely upon these technologies, the work presented in this chapter can brighten the scope of power engineers of this field.

Annotating Outline

- The upcoming era of power systems will belong to optimum utilization of FACTS and HVDC links around the globe.
- Power electronic researchers are constantly trying to improve power system performance by effective compensation.
- In this regard, the developed SVC and TCSC algorithms have been demonstrated to be quite useful in sustaining the voltage profile of the system in stressed as well as contingent states of the network (Tables 4.2 and 4.3, Figures 4.15 and 4.16).
- For faster prediction of firing angle of SVC an ANN-based expert system has also been developed to improve the dynamic response of the device (Tables 4.7 and 4.8).
- To improve the voltage and loss profile of a contingent power network (Tables 4.14 and 4.15), an HVDC interconnection link has also been modelled to assist the system operator to select the proper method of compensation considering topological and economic constraints.

- For contingency surveillance a sensitivity index has also been developed and implemented with HVDC and TCSC for congestion relief of the power network.
- Advanced controlled technologies such as FACTS devices and HVDC are quite crucial to be incorporated in the present-day power network. These technologies ask for high investment cost, but hence a deregulated power network has been developed to endorse cost-effective quality power to consumer; its optimized employment is highly preferred in power industries.

5

Multi-Objective Optimization Algorithms for Deregulated Power Market

Power engineers are constantly trying to develop multi-dimensional objectives for power networks. But as the numbers of constraints are increasing day by day, it has become a mammoth task to reach an optimum solution with classical techniques. In an endeavour optimize system performance the work presented in this chapter has been dedicated to the search for multi-objective solution algorithms to present-day power network problems like loss of sustainability, line congestion and cost volatility with the effective utilization of expert systems like genetic algorithm (GA), particle swarm optimization (PSO) and differential evolution (DE) and by the development of indices like security sensitivity index (SSI) and overloading index (OI). Deregulated power networks are concerned with producing cost-effective power; thus, any kind of network perturbation has to be treated as an economic perturbation. Hence, different perturbation measurement indices such as value of lost load (VOLL), value of congestion cost (VOCC) and value of excess loss (VOEL) have been developed in this chapter to expand a multi-objective cost-constrained deregulated power network.

5.1 Introduction

Electric utilities have been in business for more than a century. Starting from very small networks, utilities have become widespread and complex in nature. Power system utilities have gone through various stages of evolution since then. But they have been going through some unprecedented business changes over the last few years. Despite these changes, the elementary units of an electric power system remain (1) generation, (2) transmission and (3) distribution. With these three functional units, the operation of a power system is a set of complex activities that depends not only on the state of existing technology but also on other complex issues, like economy, social advancement and environmental impact. These factors, however, vary from country to country, and so do the power networks and their mode of operations. Every power generating installation, in general, involves an enormous amount of investment. That is why any change in the grid network or its operation model often raises passionate debates. But due to demand of social and technological advancements, the changes in power networks and their mode of operation are inevitable. Power systems in the early days were developed on the concept of natural monopoly. In a monopoly system, only one utility controls all three functional zones. This conventional setup is known as a vertically integrated system (VIS). The possibility of a new competitive environment in this century-old regulated power industry has created enthusiasm among researchers. The deregulation of the electric power industry involves technical and also various non-technical and economic issues. Along with power researchers, experts from economy, finance, risk management and marketing are contributing to the process of development of a deregulated power network that ensures fair competition.

Keeping these in view, this chapter presents multi-objective optimization algorithms for utility maximization of deregulated power networks. The works depicted, in the beginning, focus on the modification of conventional cost and power loss minimization and continue to search for an optimal multi-dimensional algorithm for the possible solutions of present-day power network problems.

5.2 Deregulated Power Market Structure

In a conventional system, one utility or company has the exclusive right of marketing electricity in one designated service area. The schematic diagram of a conventional system is shown in Figure 5.1.

Deregulation, on the contrary, was undertaken by introducing commercial incentives in generation, transmission and distribution of electricity. The main objective of deregulation is to achieve clear separation between production and sale of electricity and network operations. The erstwhile VIS operation has been separated into independent activities. The generation companies sell energy through competitive long-term contracts with customers or by bidding for short-term energy supply at the spot price.

On the other hand, with its significant levels of economy of scale, it was natural for the transmission sector to become a monopoly. It was therefore necessary to introduce regulation in transmission so as to prevent overcharging for its services; consequently, the transmission grid has to be a neutral monopoly subject to regulation by public authorities. A new regulatory framework

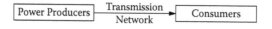

FIGURE 5.1
Schematic diagram of a conventional power network.

has been established to offer third parties 'open access' to the transmission network so as to overcome the monopolistic characteristics of transmission.

Due to socio-economic boundaries, it was not possible to enforce open access that is deregulation overnight. The power networks around the globe are under a continuous process of transformation from monopolistic to privatization. Though the open-access market should not have a common regulatory body, a neutral regulatory operator is required for smooth operation and control of different kinds of markets with their participants. Hence, in this deregulated environment, a system operator (SO) is assigned to perform the central coordination role with the responsibility of keeping the system in balance, i.e. to ensure that the production and imports continuously match the consumption and exports. It is required to be an independent authority without any involvement in the market competition or owning any generation facility for business (except some emergency use). Hence, it is renowned as an independent system operator (ISO).

Figure 5.2 shows a typical structure of a deregulated power network with the complex interactions among different sectors in the system.

Among the countries where the electricity supply industry has been deregulated, the South American countries, including Chile, Argentina, Bolivia,

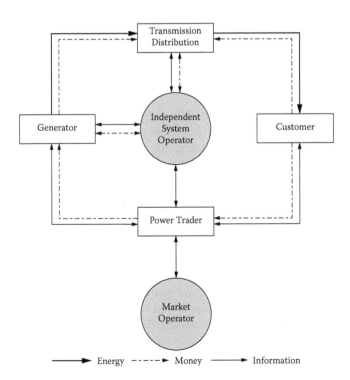

FIGURE 5.2
Typical structure of a deregulated power network.

Colombia, Peru and Brazil, were the initiators of deregulation as early as 1982. The United Kingdom, Norway, Australia, New Zealand, United States, Sweden and other European countries subsequently opened up their power sectors to competition in the 1990s. It should, however, be noted that the form of deregulation differs in each country and even among the various systems in the industries.

This deregulation is, however, a dynamic concept and brings radical changes in the power system area. Under deregulation, transmission will be controlled by ISO, a neutral organization that would provide equal access rights to all interested generating utilities. Unlike in the monopoly system, generation utilities will no longer have ownership or control of transmission or distribution facilities. The role of generating utilities would be restricted to selling power only. They can sell their power through bilateral contracts or the power pool. In a deregulated environment, various generation utilities would compete with each other for selling their product. Thus, in a deregulated system, an ISO will play the role of a supervisor for system planning and security. An ISO should execute the following duties:

- Planning services
- Power market administrative services
- Operations planning (scheduling) services
- Real-time operations (dispatch) services
- Metering, settlement and billing services
- Open information communication services

In case of congestion, due to any type of contingency or major load increase, the ISO would figure out customer priority based on predetermined criteria and cut off power of the least-priority customer to avoid major disaster and improve system security. An ISO will also work as a spot market for instant selling and buying of power. It would keep track of all transactions and calculate the transmission usage for each generator or power producer. Once more, in normal as well as contingent states, transmission loss occurs in a system as a result of flow of power through its transmission network. An ISO may also be responsible for allocation and management of the stated transmission loss. It should be able to calculate and buy an optimum amount of power from the spot market to make up for this transmission loss. Consequently, transmission loss optimization is an imperative aspect for deregulated power environments even in the contingent state.

Hence, in a deregulated power market, the ISO has to continuously monitor three different aspects: the transmission congestion, transmission losses and generation cost during normal as well as contingent states. For social welfare and the benefit of the market participants, these three parameters are to be minimized. The forecasted spot price will determine the amount of the stated welfare considering the stated parameters.

As the three features are interrelated, their simultaneous optimization needs a complex formulation of an optimization problem, and the constraints to be handled may go beyond the capacity of the classical optimization methods. Classical methods of optimization with a large number of constraints may converge to local optima. The necessity of the emerging evolutionary optimization techniques, in this context, can appear to be efficient, and their implementation for the problems of a power system may generate comparatively better solutions than classical techniques of optimization.

5.3 Soft Computing Methodologies for Power Network Optimizations

The aim of optimization is to determine the best-suited solution to a problem under a given set of constraints. Several researchers over the decades have come up with different solutions to linear and non-linear optimization problems. Mathematically, an optimization problem involves a fitness function describing the problem, under a set of constraints representing the solution space for the problem. Unfortunately, most of the traditional optimization techniques are centred around evaluating the first derivatives to locate the optima on a given constrained surface. Because of the difficulties in evaluating the first derivatives to locate the optima for many rough and discontinuous optimization surfaces, in recent times, several derivative-free optimization algorithms have emerged. The optimization problem is now represented as an intelligent search problem, where one or more agents are employed to determine the optima on a search landscape, representing the constrained surface for optimization problems [295].

In the later quarter of the 20th century, Holland [296] pioneered a new concept of evolutionary search algorithms and came up with a solution to the so far open-ended problem of non-linear optimization. Inspired by the natural adaptations of the biological spices, Holland echoed the Darwinian theory through his most popular and well-known algorithms, currently known as the genetic algorithm (GA) [296]. Holland and his co-workers, including Goldberg and Dejong, popularized the theory of GA and demonstrated how biological crossovers and mutation of chromosomes can be realized in the algorithm to improve the quality of the solution over successive iterations. In the mid-1990s Eberhart and Kennedy enunciated an alternative solution to the complex non-linear optimization problem by emulating the collective behaviour of bird flocks, particles, the boids method of Craig Reynolds and socio-cognition and called the brainchild particle swarm optimization (PSO). Around the same time, Price and Storn made a serious attempt to replace the classical crossover and mutation operator in GA by alternative operators, and consequently came up with a suitable differential operator to handle the problem. They proposed a new algorithm based on this operator and called it differential evolution (DE) [297].

All the above-mentioned algorithms do not require any gradient information of the function to be optimized, use only primitive mathematical operators and are conceptually very simple. They can be implemented in any computer language very easily and require minimal parameter tuning. Algorithm performance does not deteriorate severely with the growth of the search space dimensions as well. These issues perhaps have a great role in the popularity of these algorithms within the domain of machine intelligence and cybernetics. Table 5.1 shows the comparison of different evolutionary algorithms as applies to optimization problems in power networks. The following sub-sections present detailed descriptions of these soft computing algorithms along with their domain of applications.

5.3.1 Overview of Genetic Algorithm

Generally speaking, genetic algorithms (GAs) are simulations of evolution, of whatever kind. In most cases, however, GAs are nothing else than probabilistic optimization methods, which are based on the principles of evolution. This idea appears first in 1967 in J. D. Bagley's thesis, 'The Behavior of Adaptive Systems Which Employ Genetic and Correlative Algorithms' [298]. The theory and applicability was then strongly influenced by J. H. Holland, who can be considered the pioneer of genetic algorithms [299]. Since then, this field has witnessed tremendous development.

The genetic algorithm starts with an initial set of random solutions, the populations. Each individual in the populations is a chromosome, representing a solution to the problem. A chromosome is a string structure, typically a concentrated list of binary digits representing a coding of the control parameters of a given problem. The chromosome evolves through successive iteration called generations. The chromosomes of each generation are evaluated using some measure of fitness. To create the next generation, new chromosomes, called offspring, are formed by either merging two chromosomes from the current generation using a crossover operator or modifying chromosomes using a mutation operator. A new generation is formed by selecting, according to the fitness value, some of the parents and offspring, and rejecting others in order to keep the population size constant. Suitable chromosomes having higher survival probabilities are selected. Normally the roulette wheel selection approach is selected. After several generations, the algorithm converges to the best chromosome, which hopefully represents the optimal or near-optimal solution to the problem.

For proper understanding, GAs working with a fixed number of binary strings of fixed length have been represented here.

For this purpose, let us assume that the considered strings are all from the set

$$S = \{0,1\}^n \tag{5.1}$$

TABLE 5.1

Comparison of Evolutionary Algorithms as Applied to Optimization Problems in Power Networks

Criteria	Genetic Algorithm	Particle Swarm Optimization	Differential Evolution
Generation of solution variables	Random and within the region bounded by constraints	Random and within the region bounded by constraints	Random but high precision within the region bounded by constraints
Evaluation of fitness	At every step up to given number of iterations	At every step up to given number of iterations	At every step up to given number of iterations
Updating of population	By crossover and mutation but requires low mutation rate for optimization	Particles update themselves from internal velocity	By crossover and mutation but it takes correlated self-adopting mutation step sizes in order to make timely progressive optimization
Solution memory	Chromosomes share information with each other so the whole population moves toward the solution; hence requires huge memory support	Each particle has memory and only g_{best} gives information to others	Solutions do not have memory; hence relies on information sharing at every step
Constraint handling	Low mutation rate may get troubled with problems having interdependent constraints	As the particles keep track on p_{best} and g_{best}, interdependent constraints can be easily handled	Due to self-adopting mutation, it can handle interdependent constraint variables
Quality of optimization	May be premature and can lead to local solution	Generally gives global optima	Solutions are generally global
Convergence speed	Lowest dependency on the mutation rate	Very high dependency on the number of constraints	Low dependency on the number of constraints
Special feature	Can optimize convex and non-linear function	Can also optimize non-linear function with an advantage that its behaviour is convergent with oscillation and zigzagging	Can optimize non-linear and non-differentiable continuous space functions with real-valued parameters

where n is obviously the length of the strings. The population size will be denoted with m in the following. Therefore, the generation at time t is a list of m strings, which can be denoted as follows:

$$B_t = (b_{1,t}, b_{2,t}, \ldots \ldots b_{m,t}) \tag{5.2}$$

Genetic algorithms obey the following structure:

Algorithm

```
t: = 0;
Compute initial population B₀ = (b₁,₀,b₂,₀,........bₘ,₀)
WHILE stopping condition not fulfilled DO
BEGIN
FOR i: = 1 to m DO
Select an individual bᵢ,ₜ₊₁ from Bₜ
FOR i: = 1 to m-1 STEP 2 DO
IF Random[0, 1] ≤ pc THEN
Cross bᵢ,ₜ₊₁ with bᵢ₊₁,ₜ₊₁;
FOR i: = 1 to m DO
Eventually mutate bᵢ,ₜ₊₁;
t: = t + 1;
END
```

Obviously, selection, crossover (done only with a probability of pc here) and mutation are still degrees of freedom, while the sampling operation is already specified. As it is easy to see, every selected individual is replaced by one of its children after crossover and mutation; unselected individuals die immediately. This is a rather common sampling operation, although other variants are known and reasonable.

Selection is the component that guides the algorithm to the solution by preferring individuals with high fitness to low-fitted ones. It can be a deterministic operation, but in most implementations it has random components. One variant, which is very popular nowadays, is the following scheme, where the probability to choose a certain individual is proportional to its fitness. It can be regarded as a random experiment with

$$P[b_{j,t}] = \frac{f(b_{j,t})}{\sum_{k=1}^{m} f(b_{k,t})} \tag{5.3}$$

Of course, this formula only makes sense if all the fitness values are positive. If this is not the case, a non-decreasing transformation, $\varphi: R \rightarrow R^+$, must be applied (a shift in the simplest case). Then the probabilities can be expressed as

$$P[b_{j,t}] = \frac{\varphi f(b_{j,t})}{\displaystyle\sum_{k=1}^{m} \varphi f(b_{k,t})} \tag{5.4}$$

The algorithmic formulation of the selection scheme can be written down as follows:

Algorithm

```
x:= Random[0,1];
i:= 1
```

$$\textbf{WHILE } i < m \ \& \ x < \sum_{j=1}^{i} f(b_{j,t}) / \sum_{j=1}^{m} f(b_{j,t}) \textbf{ DO}$$

```
i:= i + 1
Select b_{i,t};
```

For obvious reasons, this method is often called proportional selection.

In the following step, crossover is the exchange of genes between the chromosomes of the two parents. In the simplest case, this process can be realized by cutting two strings at a randomly chosen position and swapping the two tails. This process, which is called a one-point crossover in the following, is visualized.

One-point crossover (Figure 5.3) is a simple and often used method for GAs, which operate on binary strings. For other problems or different coding, other crossover methods can be useful or even necessary.

Algorithm

```
pos: = Random {1,.....,n−1};
FOR i: = 1 to pos DO
BEGIN
Child₁[i]: = Parent₁[i];
Child₂[i]: = Parent₂[i];
END
FOR i: = pos + 1 to n DO
BEGIN
Child₁[i]: = Parent₂[i];
Child₂[i]: = Parent₁[i];
END
```

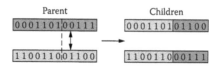

FIGURE 5.3
One-point crossover of binary string.

The last ingredient of a simple genetic algorithm is mutation – the random deformation of the genetic information of an individual by means of radioactive radiation or other environmental influences. In real reproduction, the probability that a certain gene is mutated is almost equal for all genes. So, it is convenient to use the following mutation technique for a given binary string s, where p_M is the probability that a single gene is modified:

Algorithm

```
FOR i: = 1 to n DO
IF Random [0,1]< pM THEN
invert S[i];
```

Of course, p_M should be rather low in order to avoid the GA behaving chaotically like a random search.

5.3.2 Overview of Particle Swarm Optimization

Kennedy and Eberhart introduced the concept of function optimization by means of a particle swarm [300]. Suppose the global optimum of an n-dimensional function is to be located. The function may be mathematically represented as

$$f(x_1, x_2, x_3, ...x_n) = f(\vec{X}) \tag{5.5}$$

where $\vec{x}$ is the search variable vector, which actually represents the set of independent variables of the given function. The task is to find out such an $\vec{x}$, that the function $f(\vec{x})$ value is either a minimum or a maximum denoted f^* in the search range. If the components of $\vec{x}$ assume real values, then the task is to locate a particular point in the n-dimensional hyperspace, which is a continuum of such points.

PSO is a multi-agent parallel search technique. Particles are conceptual entities, which fly through the multi-dimensional search space. At any particular instant, each particle has a position and a velocity. The position vector of a particle with respect to the origin of the search space represents a trial solution of the search problem. At the beginning, a population of particles is initialized with random positions marked by vectors $\bar{x}_i$ and random velocities $\bar{v}_i$. The population of such particles is called a swarm, S. A neighbourhood relation, N, is defined in the swarm.

N determines for any two particles P_i and P_j whether they are neighbours or not. Thus, for any particle P, a neighbourhood can be assigned as $N(P)$, containing all the neighbours of that particle. Different neighbourhood topologies and their effect on the swarm performance will be discussed later. However, a popular version of PSO uses $N = S$ for each particle. In this case, any particle has all the remaining particles in the swarm in its neighbourhood. PSO dynamics is illustrated in Figure 5.4.

Each particle P has two state variables, viz. its current position $\overline{x(t)}$ and its current velocity $\overline{v(t)}$. It is also equipped with a small memory comprising its previous best position (one yielding the highest value of the fitness function found so far), $\bar{p}(t)$, i.e. personal best experience and the best $\bar{p}(t)$ of all, $P \in N(P); \bar{g}(t)$, i.e. the best position found so far in the neighbourhood of the particle. When we set $N(P) = S, \bar{g}(t)$, is referred to as the globally best particle in the entire swarm. The PSO scheme has the following algorithmic parameters:

V_{max} or maximum velocity, which restricts $\vec{V}_i(t)$ within the interval $[-V_{max}, V_{max}]$

An inertial weight factor ω

Two uniformly distributed random numbers φ_1 and φ_2 that respectively determine the influence of $\bar{p}(t)$ and $\bar{g}(t)$ on the velocity update formula

Two constant multiplier terms C_1 and C_2 known as self-confidence and swarm confidence, respectively

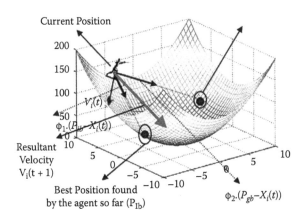

FIGURE 5.4
Illustrating the dynamics of a particle in PSO.

Initially the settings for $\vec{p}(t)$ and $\vec{g}(t)$ are $\vec{p}(0) = \vec{g}(0) = \vec{x}(0)$ for all particles. Once the particles are all initialized, an iterative optimization process begins, where the positions and velocities of all the particles are altered by the following recursive equations. The equations are presented for the dth dimension of the position and velocity of the ith particle.

$$V_{id}(t+1) = \omega.v_{id}(t) + C_1.\varphi_1.\left(P_{id}(t) - x_{id}(t)\right) + C_2.\varphi_2.\left(g_{id}(t) - x_{id}(t)\right) \quad (5.6)$$

$$x_{id}(t+1) = x_{id}(t) + v_{id}(t+1) \quad (5.7)$$

The first term in the velocity updating formula represents the inertial velocity of the particle. ω is called the inertia factor. Shi and Eberhart [297] stated C_1 as self-confidence and C_2 as swarm-confidence. These terminologies provide an insight from a sociological standpoint. Since the coefficient C_1 has a contribution to the self-exploration (or experience) of a particle, it can be regarded as the particle's self-confidence. On the other hand, the coefficient C_2 has a contribution to motion of the particles in the global direction, which takes into account the motion of all the particles in the preceding program iterations; naturally, its definition as swarm-confidence is apparent. φ_1 and φ_2 stand for a uniformly distributed random number in the interval [0, 1]. After having calculated the velocities and position for the next time step (t+1), the first iteration of the algorithm is completed. Typically, this process is iterated for a certain number of time steps, or until some acceptable solution has been found by the algorithm or an upper limit of CPU usage has been reached. The algorithm can be summarized in the following pseudo-code [301]:

PSO Algorithm

```
INPUT: Randomly initialized position and velocity of the
particles: X⃗ᵢ(0) and V⃗ᵢ(0)
OUTPUT: Position of the approximate global optima X⃗*
BEGIN
WHILE terminating condition is not reached Do
BEGIN
FOR i = 1 to number of particles
Evaluate the fitness: = f(X⃗ᵢ);
Update p⃗ᵢ and g⃗ᵢ;
Adapt velocity of the particle using equations (1);
Update the position of the particle;
INCREASE i
END WHILE
END
```

5.3.3 Overview of Differential Evolution

Like any other evolutionary algorithm, DE also starts with a population of D-dimensional search variable vectors. We will represent subsequent generations in DE by discrete time steps like $t = 0, 1, 2, ..., t, t + 1$, etc. Since the vectors are likely to be changed over different generations, we may adopt the following notation for representing the ith vector of the population at the current generation (i.e. at time $t = t$) as

$$\overrightarrow{X_i}(t) = \left[x_{i,1}(t), x_{i,2}(t), x_{i,3}(t).....x_{i,D}(t) \right] \qquad (5.8)$$

These vectors are referred to in the literature as genomes or chromosomes. DE is a very simple evolutionary algorithm. For each search variable, there may be a certain range within which the value of the parameter should lie for better search results. At the very beginning of a DE run or at $t = 0$, problem parameters or independent variables are initialized somewhere in their feasible numerical range. Therefore, if the jth parameter of the given problem has its lower and upper bounds as x_j^L and x_j^U, respectively, then we may initialize the jth component of the ith population members as

$$x_{i,j}(0) = x_j^L + rand(0,1).\left(x_j^U - x_j^L\right) \qquad (5.9)$$

where $rand(0, 1)$ is a uniformly distributed random number lying between 0 and 1. Now in each generation (or one iteration of the algorithm), to change each population member $\overrightarrow{X_i}(t)$, for example, a donor vector $\overrightarrow{V_i}(t)$ is created. It is the method of creating this donor vector that demarcates between the various DE schemes. However, here one such specific mutation strategy known as DE/rand/1 has been discussed. In this scheme, to create $\overrightarrow{V_i}(t)$ for each i_{th} member, three other parameter vectors (say the r_1, r_2 and $r_3{}^{th}$ vectors) are chosen in a random fashion from the current population. Next, a scalar number F scales the difference of any two of the three vectors, and the scaled difference is added to the third one to obtain the donor vector $\overrightarrow{V_i}(t)$. The process for the jth component of each vector can be expressed as

$$v_{i,j}(t+1) = x_{r1,j}(t) + F.\left(x_{r2,j}(t) - x_{r3,j}(t)\right) \qquad (5.10)$$

The process is illustrated in Figure 5.5. Closed curves in Figure 5.5 denote constant cost contours; i.e. for a given cost function f, a contour corresponds to $f(\overrightarrow{X}) = $ constant. Here the constant cost contours are drawn for the Ackley function.

Next, to increase the potential diversity of the population, a crossover scheme comes in to play. DE can use two kinds of crossover schemes: exponential and binomial. The donor vector exchanges its 'body parts',

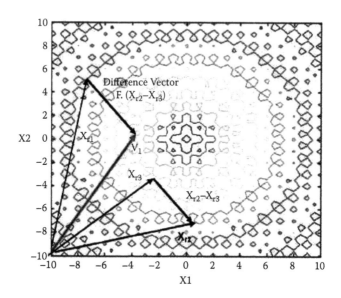

FIGURE 5.5
Illustrating creation of the donor vector in two-dimensional parameter space.

i.e. components with the target vector $\overline{X}_i(t)$ under this scheme. In exponential crossover, an integer n has been chosen randomly among the numbers [0,D−1]. This integer acts as a starting point in the target vector, from where the crossover or exchange of components with the donor vector starts. We also choose another integer, L, from the interval [1,D]. L denotes the number of components; the donor vector actually contributes to the target. After a choice of n and L, the trial vector

$$\overline{U}_i(t)=\left[u_{i,1}(t),u_{i,2}(t),.....u_{i,D}(t)\right] \tag{5.11}$$

is formed with

$$u_{i,j}(t) = v_{i,j}(t) \text{ for } j = \langle n \rangle_D, \langle n+1 \rangle_D \text{ ...... } \langle n-L+1 \rangle_D$$
$$= x_{i,j}(t) \tag{5.12}$$

where the angular brackets $\langle \ \rangle_D$ denote a modulo function with modulus D. The integer L is drawn from [1,D] according to the following pseudo-code:

```
L = 0;
DO
{
L = L + 1
} WHILE (rand (0,1)<CR) AND (L>D);
```

Hence, in effect probability $(L>m) = (CR)^{m-1}$ for any $m > 0$. CR is called crossover constant, and it appears as a control parameter of DE just like F. For each donor vector V, a new set of n and L must be chosen randomly as shown above. However, in the binomial crossover scheme, the crossover is performed on each of the D variables whenever a randomly picked number between 0 and 1 is within the CR value. The scheme may be outlined as

$$u_{i,j}(t) = v_{i,j}(t) \text{ if } rand\,(0,1) < CR$$
$$= x_{i,j}(t) \text{ else} \qquad (5.13)$$

In this way for each trial vector $\overline{X_i}(t)$ an offspring vector $\overline{U_i}(t)$ is created. To keep the population size constant over subsequent generations, the next step of the algorithm calls for selection to determine which one of the target vector and the trial vector will survive in the next generation, i.e. at time $t = t + 1$. DE actually involves the Darwinian principle of survival of the fittest in its selection process, which may be outlined as

$$\overline{X_i}(t+1) = \overline{U_i}(t) \text{ if } f\left(\overline{U_i}(t)\right) \leq f\left(\overline{X_i}(t)\right)$$

$$= \overline{X_i}(t) \text{ if } f\left(\overline{X_i}(t)\right) < f\left(\overline{U_i}(t)\right) \qquad (5.14)$$

where $f(\,)$ is the function to be minimized. So if the new trial vector yields a better value of the fitness function, it replaces its target in the next generation; otherwise, the target vector is retained in the population. Hence, the population either gets better (with regard to the fitness function) or remains constant, but it never deteriorates. The DE/rand/1 algorithm is outlined below:

Procedure of DE

```
INPUT: Randomly initialized position and velocity of the
particles: X⃗ᵢ (0)
OUTPUT: Position of the approximate global optima X⃗*
BEGIN
Initialize population;
Evaluate fitness;
FOR i = 0 to max-iteration DO
BEGIN
Create Difference-Offspring;
Evaluate fitness;
IF an offspring is better than its parent
THEN replace the parent by offspring in the next generation;
END IF;
END FOR;
END.
```

The choice of GA, PSO and DE is highly influenced by the type and degree of non-linearity of the projected problem. GA can create a new solution with the help of crossover and the mutation technique, but if the solution lies within the initial population, then these processes become irrelevant. PSO, on the other hand, chooses the best solution from the initial population, but as it cannot create a new solution, it suffers from a social metaphor. DE, like GA, also uses crossover and mutation for developing a new solution, but it limits the time complexity with the help of a differential operator. In general, for a solution without time, bounded GA may be chosen; for fast and accurate results, PSO can be adopted; and for a fast, accurate and global solution, DE can be employed.

In this chapter, the above-discussed three different soft computing techniques (genetic algorithm, particle swarm optimization and differential evolution) have been used for the development of new optimization methodologies to conquer different power network problems maintaining the consumer welfare. But, before that, a comparative study of these techniques has been portrayed here in tabular form for ready reference.

5.4 Algorithms for Utility Optimization with Cost and Operational Constraints

Utility optimization without violating the operational constraints has been the focus of researchers since the birth of electrical power networks. The objective is overall cost minimization with a feasible generation schedule, considering the equality and inequality constraints. Several cost optimization models have been proposed over the years, leading to utility maximization. As the power networks grew large, the transmission efficiency became a great concern, and the researchers started to develop power loss minimization models.

In the following sub-sections some novel methodologies of power loss and generation cost minimization have been described. These algorithms have been tested on a standard system and the results were quite promising.

5.4.1 Genetic Algorithm-Based Cost-Constrained Transmission Line Loss Optimization

In this section, a systematic method for transmission loss optimization using GA and power flow analysis for deregulated power markets has been projected [308]. The basic concept of this loss optimization is laid under the optimal power flow (OPF) model, where line loss is a function of the B coefficient of active power and the C coefficient of reactive power. The results of this loss optimization have been coupled with load flow to check the overall operating loss of the system. The methodology developed tries to minimize the operating loss of the system by optimizing internal parameters such as tap position of the existing transformer in the network. For comparison of the results, the conventional cost optimization method has been adopted.

Objective function for conventional cost optimization (method 1):

$$F(x) = \text{minimize} \sum_{i \in NG} C_i(P_{Gi}) = \sum_{i=1}^{NG} \alpha_i \left(P_{G_i}\right)^2 + \beta_i P_{G_i} + \gamma_i \tag{5.15}$$

Objective function of the developed line loss optimization (method 2):

$$F(x) = \text{minimize}$$

$$(P_L) = \sum_{i=1}^{n} \sum_{j=1}^{m} P_{G_i} B_{ij} P_{G_j} = B_{00} + \sum_{i=1}^{n} B_{i0} P_{G_i} + \sum_{i=1}^{n} \sum_{j=1}^{m} P_{G_i} B_{ij} P_{G_j} \tag{5.16}$$

$$F(x) = F'(x) = \text{minimize}$$

$$(Q_L) = \sum_{i=1}^{n} \sum_{j=1}^{m} Q_{G_i} C_{ij} Q_{G_j}$$

$$= C_{00} + \sum_{i=1}^{n} C_{i0} Q_{G_i} + \sum_{i=1}^{n} \sum_{j=1}^{m} Q_{G_i} C_{ij} Q_{G_j} \tag{5.17}$$

P_L and Q_L are the active and reactive loss terms and can be expressed using B and C coefficients as follows:

$$P_L = \sum_{i=1}^{n} \sum_{j=1}^{m} P_{G_i} B_{ij} P_{G_j} = B_{00} + \sum_{i=1}^{n} B_{i0} P_{G_i} + \sum_{i=1}^{n} \sum_{j=1}^{m} P_{G_i} B_{ij} P_{G_j} \tag{5.18}$$

where

$$B_{ij} = \frac{\cos(\theta_i - \theta_j) R_{ij}}{\cos \varphi_i \cos \varphi_j |V_i| |V_j|} \qquad B_{i0} = -\sum_{j=1}^{m} (B_{ij} + B_{ji}) P_{D_j} \quad \text{and}$$

$$B_{00} = \sum_{i=1}^{n} \sum_{j=1}^{m} P_{D_i} B_{ij} P_{D_j} \tag{5.19}$$

$$Q_L = \sum_{i=1}^{n} \sum_{j=1}^{m} Q_{G_i} C_{ij} Q_{G_j} = C_{00} + \sum_{i=1}^{n} C_{i0} Q_{G_i} + \sum_{i=1}^{n} \sum_{j=1}^{m} Q_{G_i} C_{ij} Q_{G_j} \tag{5.20}$$

$$C_{ij} = \frac{\cos(\theta_i - \theta_j)X_{ij}}{\cos\varphi_i \cos\varphi_j \,|V_i|\,|V_j|} \quad C_{i0} = -\sum_{j=1}^{m}(C_{ij} + C_{ji})Q_{D_j} \text{ and}$$

$$C_{00} = \sum_{i=1}^{n}\sum_{j=1}^{m} Q_{D_i} C_{ij} Q_{D_j}$$

where

$$\theta_j = \delta_j - \varphi_j \quad \text{and} \quad \theta_j = \delta_j - \varphi_j \tag{5.21}$$

The flowchart of the developed methodology has been shown in Figure 5.6.

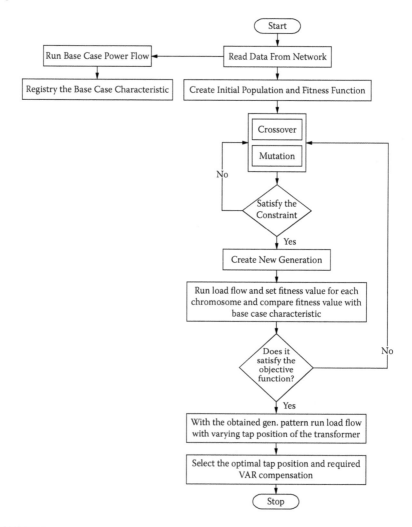

FIGURE 5.6
Flowchart of the solutions methodology.

At the beginning of the GA optimization process, each variable gets a random value from its predefined domain. The generator power outputs have well-defined lower and upper limits, and the initialization procedure commences with these limits given by

$$P_{G_i}^{min} \leq P_{G_i} \leq P_{G_i}^{max} \text{ and } Q_{G_i}^{min} \leq Q_{G_i} \leq Q_{G_i}^{max} \tag{5.22}$$

In the next step, the implementation of a problem in GA is realized within the constraint function. The constraints of power balance can be written as follows:

$$\varepsilon_1 = \sum_{i=1}^{n} P_{G_i} - P_D - P_L \text{ and } \varepsilon_2 = \sum_{i=1}^{n} Q_{G_i} - Q_D - Q_L \tag{5.23}$$

In the first step of this methodology, the optimized values of generated power for all generators have been determined by GA. Table 5.2 illustrates the comparison between the solutions obtained by conventional cost optimization (method 1) and developed line loss optimization (method 2) for the IEEE 30 bus test system (described in Appendix A).

With the generations obtained from GA, load flow analysis has been executed to get total line loss for different tap positions of the regulating transformer. It can be observed that with different tap positions, the active and reactive power flows do not remain the same. Actually, the on-load

TABLE 5.2

Comparison of Generator Contributions Obtained from Conventional Cost Optimization Method (Method 1) and Developed Method (Method 2)

Contribution of Generators		Method 1	Method 2
GENCO 1	P_G (p.u.)	1.384	1.235
Bus No. 1	Q_G (p.u.)	−0.185	−0.02
GENCO 2	P_G (p.u.)	0.575	0.682
Bus No. 3	Q_G (p.u.)	−0.0056	−0.0065
GENCO 3	P_G (p.u.)	0.245	0.339
Bus No. 5	Q_G (p.u.)	0.212	0.204
GENCO 4	P_G (p.u.)	0.35	0.334
Bus No. 8	Q_G (p.u.)	0.267	0.254
GENCO 5	P_G (p.u.)	0.179	0.105
Bus No. 11	Q_G (p.u.)	0.241	0.247
GENCO 6	P_G (p.u.)	0.169	0.207
Bus No. 13	Q_G (p.u.)	0.317	0.318
Total system loss	P_L (p.u.)	0.074	0.067
	Q_L (p.u.)	−0.257	−0.447

TABLE 5.3

Load Flow Study with GA-Generated OPF Result

Sl No.	Tap Position	Operating Line Loss with Regard to Optimized Line Loss		Generation of Fictitious Generator	
		Active p.u.	Reactive p.u.	Active p.u.	Reactive p.u.
1.	Normal	0.998	1.840	1.233	−0.00983
2.	2.5% increase	0.976	1.820	1.231	−0.02396
3.	5% increase	0.970	1.640	1.231	−0.03764
4.	7.5% increase	0.979	1.270	1.232	−0.05089
5.	2.5% decrease	1.030	1.686	1.235	0.00475
6.	5% decrease	1.035	1.364	1.238	0.01979
7.	7.5% decrease	1.040	0.878	1.243	0.03530

tap-changing transformers regulate nodal voltage magnitude by varying the transformer tap position. Due to alteration of voltages, line flow changes. Thus, for optimized line loss, optimal tap position can be calculated. From Table 5.3, it can be inferred that the effect on active loss due to change of tap position of the transformer is minor; rather, it can be considered optimized. But there is remarkable change in line loss due to reactive power flow with the alteration of tap positions of the transformer.

In the condition (sl no. 1–4 of Table 5.3), a fictitious generator has been operated within its limits. It can be observed that at this stage, no compensation is required for line loss optimization. But the reactive power requirement for reactive loss optimization with different combinations of decreasing tap position (sl no. 5–7 of Table 5.3) may exceed its limits. For this uncovered reactive power, the developed methodology investigates the need for a reactive power generator in the system.

Thus, this methodology can be implemented with compensating devices for further minimization of loss; however, that compensation can lead to an additional investment cost. But this technique can assist the ISO to settle the need for compensation in the system and also to prepare a generation schedule by optimizing loss in the system. The ISO in this scenario has to choose the accurate tap position to optimize loss in deregulated power networks.

5.4.2 GA-Based Generation Cost-Constrained Redispatching Schedules of GENCOs

As a deregulated power market is intended to open the power sector to market forces with the ultimate target of reducing consumer prices, the central ideology of electric power industry deregulation remains to decouple the delivery of power from the purchase of the power itself, and for it to be priced and contracted separately. From an economic point of view, active power pricing in a competitive power market in variable loading conditions presents a good potential for providing valuable instructions for system operations. A novel algorithm has been designed for optimal allocation of a generation schedule of generators to optimize generation cost under stressed conditions of a system considering consumer welfare [314]. Here, the optimal generation dispatch problem is formulated as a non-linear-constrained optimization problem where real power generation and total generation cost are to be optimized simultaneously. This algorithm is utilized for the optimal allocation of generated power to minimize the price of electricity. Here the formulation of the optimal value of generation cost and generated power can be expressed as follows:

$$\min C_{TOTAL} = C_1 (P_G) \tag{5.24}$$

subject to

$$E(g) = 0, B(g) \le b \tag{5.25}$$

$C_1 (P_G)$, the total generation cost, is defined as

$$C_{total} \left(= \sum_{i=1}^{NG} C_i \right) = \sum_{i=1}^{NG} \alpha_i \left(P_{G_i} \right)^2 + \beta_i P_{G_i} + \gamma_i \tag{5.26}$$

where $E(g)$ is equality constraint, $B(g)$ is inequality constraint and P_G is power generation of generators.

After encoding, the objective function (fitness) will be evaluated for each individual of the population. In this work, the fitness is defined as follows:

$$Fitness = S - C_{TOTAL} \tag{5.27}$$

Here S is the forecasted common bidding price of consumer and supplier. A constraints function is achieved using the conventional power balance relation as described in (5.22). Table 5.4 illustrates the solution obtained by GA for total generation cost and optimum percentage loading of each generator.

Combinations B to G of Table 5.5 describe the different redispatching schedules where a particular generator did not take part in delivering additional power during stressed conditions, i.e. in combination B the generation

TABLE 5.4

Optimal Percentage Loading of Each Generator and Total Generation Cost

Generator Loading (in percent of maximum capacity) (generation in MW)						Generation Cost US$/h
Gen 1	Gen 2	Gen 3	Gen 4	Gen 5	Gen 6	
82.82	71.67	88.17	89.08	83.7	83.7	2063.5
(120.5)	(50.6)	(31.4)	(44.54)	(21.6)	(21.7)	

TABLE 5.5

Load Allocation for 10% Increase in Demand and Generation Cost with Different Combinations

Combinations	Percentage Loading of Generators for 10% Increment in Demand						Generation Cost US$/h	Remarks
	Gen 1	Gen 2	Gen 3	Gen 4	Gen 5	Gen 6		
A	86.12	85.83	99.71	99.98	99.96	99.96	2292.5	No gen is fixed
B	82.82	92.62	99.99	99.99	99.99	99.99	2282.7	Gen 1 is fixed
C	92.98	71.67	99.99	99.99	99.99	99.99	2344	Gen 2 is fixed
D	86.12	91.76	88.17	99.99	99.99	99.99	2308.1	Gen 3 is fixed
E	87.82	89.62	99.99	89.08	99.99	99.99	2338.32	Gen 4 is fixed
F	85.92	92.16	96.26	99.99	83.77	99.99	2303.12	Gen 5 is fixed
G	86.25	91.45	99.99	99.99	99.99	83.77	2308.95	Gen 6 is fixed

of generator 1 and that of generator 2 in combination C are kept fixed at their prior generations to increase in demand and so on. Table 5.5 depicts that ISO has different options of allocating the increase in demand to generators for maintaining generation cost at its minimum possible value for 10% increment in demand. From these redispatching schedules, ISO can choose one where the generation cost is optimum. For instance, for 10% increment in demand, redispatching schedule B is the most appropriate one from the buyers' point of view.

Thus, with this methodology, ISO can maintain the generation cost within a predefined limit even in stressed conditions. Hence, the variation of electricity price can be restricted, which has an extremely significant impact on deregulated environments.

5.5 Congestion Management Methodologies

Transmission congestion occurs when there is insufficient transmission capacity to simultaneously accommodate all requests for transmission service within a region. Historically, vertically integrated utilities managed this condition by constraining the economic dispatch of generators with the objective of ensuring security and reliability of their own or neighbouring systems.

Electric power industry restructuring has moved generation investment and operation decisions into the competitive market, but has left transmission as a communal resource in the regulated environment. This mixing of competitive generation and regulated transmission makes congestion management difficult. The difficulty is compounded by the increase in the amount of congestion resulting from increased commercial transactions and the relative decline in the amount of transmission. Transmission capacity, relative to peak load, has been declining in all regions of the power network for over a decade. This decline is expected to continue.

Congestion management schemes used today have negative impacts on energy markets, such as disruptions and monetary penalties. To mitigate these concerns, various congestion management methods have been proposed, including redispatch and curtailment of scheduled energy transmission. In the restructured electric energy industry environment, new congestion management approaches are being developed that strive to achieve the desired degree of reliability while supporting competition in the bulk power market.

The first responsibility of the transmission system operator (SO), whether it is a large ISO or a small utility control area, is to maintain system reliability. This involves developing generation and load schedules that can be balanced in real time. The SO must make sure that the scheduled flows do not exceed a maximum for any link on the system. Scheduling generators and loads must carefully consider any transmission link that could potentially become constrained. This consideration includes not only the current flows on the system's lines and equipment, but also the post-contingency capacity. For any link the transmission system must provide enough capacity that any single contingency within the system (and any credible multiple contingency) could handle it.

Although the transmission system operates according to the physical laws of power flow, the economic implications for congestion management are equally important. Transmission congestion can be easily managed by curtailing loads. However, arbitrarily restricting loads can have significant economic costs. Due to line congestion the locational marginal price (LMP) may rise, and along with the congestion cost, and the buyer has to pay for this cost of extra energy transfer. Hence, line congestion introduces an unexpected rise in spot price to be forecasted for electrical power, which may infringe the ongoing transaction between the participants. The possible solutions and their relative benefits are formulated in Figure 5.7.

Among the possible solutions of line congestion management, generation rescheduling with and without load curtailment is prudent. With load curtailment the consumer loses the reliability of power and the suppliers suffer the revenue loss. But again, generation rescheduling without load curtailment will increase the spot price of electricity due to recontribution of generators. In this case, the end user suffers monetary loss. Both techniques have their own advantages and disadvantages, but one has to

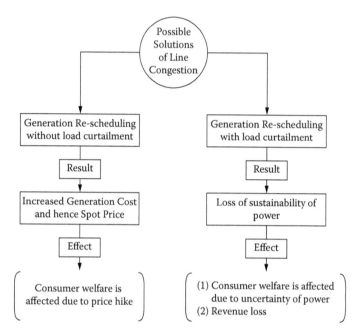

FIGURE 5.7
Comparison of different congestion management solutions.

choose from them considering the topological and socio-economic state of the system. Considering the congestion management methods, the following sub-section depicts the utility of soft computing techniques for effective line overload alleviation in deregulated power markets. The economic aspects of congestion management have also being taken care of without degrading the reliability of the system.

5.5.1 Generator Contribution-Based Congestion Management Using Multi-Objective GA

Thus, congestion management is another key function of the system operator in the restructured power industry during an unexpected contingency. A new methodology can be used for congestion management using multi-objective GA [311]. In this algorithm, the contribution of generators by both active and reactive line loss optimizations can be used to relieve the congested line, found by the developed overloading index (OI) (5.28) during contingency. The planned methodology depicts the information related to the optimization status of total system loss as well as generators' contribution schedule to minimize the investment cost without installing any external devices and to maximize the consumer welfare by avoiding any load curtailment for congestion management and without affecting the voltage profile of the system.

Through the optimization method, GA, the optimized values of rescheduled generation for all GENCOs have been determined considering all equality and inequality constraints of optimal power flow and by taking optimizations of both active and reactive line losses as objective functions (Table 5.2).

The changes in active and reactive power contribution for all GENCOs (for the IEEE 30 bus test system) (Figure 5.8) are within their specified limit as described in (5.21).

When a line is tripped by an unexpected limit violation, there is a possibility of another line overflow because the power that had flowed through the tripped line should now flow elsewhere. In these circumstances, a remedial action has to be taken to maintain system security. The severity of the line outage depends on the amount of power that had flowed through the tripped line.

To monitor the line overflow, during generation or transmission line contingency an index has been formulated and named overloading index, which can be defined as a change in power flow through a transmission line during the contingency of other lines. Mathematically it can be expressed as follows:

$$\mu_{mn} = \frac{P_{mn} - \overline{P_{mn}}}{P_{mn}} \tag{5.28}$$

where P_{mn} and $\overline{P_{mn}}$ are the active power flow through the line $m\text{-}n$ after contingency and before contingency, respectively. A higher value of this index indicates the more congested line in the power network.

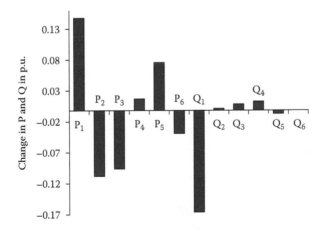

FIGURE 5.8
Change in real and reactive power generation.

In this test, five different lines have been tripped that were chosen according to the amount of line flow considered. For a particular line fault, the four most congested lines have been found. With the calculated contribution of GENCOs, congested lines can be relieved from overloading (Table 5.6). A remarkable reduction in line congestion has been observed with developed generation with line loss optimization (method 2) compared with the generation obtained from the conventional cost optimization technique (method 1). This reduction of overloading may reduce the cost of congestion, which is an integral part of operating cost of generation in deregulated environments of power systems.

The other important aspect of this developed methodology is the reduction of the system operating losses during contingency. Table 5.7 compares the losses with original contribution (result obtained from conventional cost optimization (method 1)) and recontribution of GENCOs with line loss optimization (method 2) (Table 5.2). As shown, the system operating losses have decreased by a considerable amount with the recontribution schedule of GENCOs. Hence, it can be stated that along with the reduction of congestion, this methodology can lower system operating losses.

The above advantages of this recontribution schedule remain less consequential unless it has atleast an effect on the operating conditions of the system. The voltage profiles shown in Figures 5.9 to 5.11 for the test cases strengthen the competency of the recontribution schedule compared to the normal generation schedule. During contingency the voltage profile remains the least affected by the imposed schedule. This implies that without affecting the voltage, this schedule can offer significant benefits, like minimization of system losses and prices of electricity in terms of congestion management even during contingency. The other important advantage of a calculated recontribution schedule that has been prepared by GA by optimizing the active and reactive losses in deregulated electricity markets is that it does not threaten the economic dispatch, and it also offers promising results with power flow analysis.

5.5.2 DE- and PSO-Based Cost-Governed Multi-Objective Solutions in Contingent State

In this section, the aim of this effort is power generation cost minimization such that the least variation of electricity price is encountered with respect to its precontingent value and releasing the overload on the congested line. As described in Figure 5.7, generation rescheduling with load curtailment may have disadvantages like affecting reliability and loss of revenue with reference to generation rescheduling without load curtailment, but the price in the former method may be maintained at or below its precontingent value by limiting the deviation of generation cost. This necessitates a complex formulation of optimization problem, which can be solved by efficient optimization techniques like DE and PSO.

TABLE 5.6

Congestion Management with Enumerated Generation Pattern

	Line Fault	Four Most Congested Lines	Power Flow before Fault (A)	Method 1 (B)	Method 2 (C)	% Overload $\left(\dfrac{B-A}{A}\right)*100$	% Overload with Proposed Generation $\left(\dfrac{C-A}{A}\right)*100$
				Power Flow in p.u. after Fault with Generation Using			
Case 1	2–4	2–6	0.3802	0.5277	0.4099	38.79	7.80
		3–4	0.4481	0.5552	0.4563	23.91	1.82
		1–3	0.4816	0.5937	0.5039	23.27	4.63
		2–5	0.5802	0.6379	0.5946	9.94	2.48
Case 2	2–5	2–6	0.3803	0.6666	0.4036	75.30	6.12
		2–4	0.2911	0.5043	0.3186	73.22	9.44
		12–16	0.0594	0.0837	0.0604	40.90	1.68
		24–25	0.0280	0.0316	0.0290	12.91	3.57
Case 3	6–7	8–6	0.0155	0.0255	0.0157	63.97	1.29
		6–9	0.2076	0.2489	0.2153	19.88	3.70
		9–10	0.1734	0.1973	0.1872	13.74	7.95
		6–10	0.1096	0.1214	0.1132	10.70	3.28
Case 4	12–15	14–15	0.0535	0.1032	0.0557	92.98	4.11
		12–16	0.0594	0.1000	0.0692	68.19	16.4
		22–24	0.0943	0.1237	0.1074	31.16	13.8
		4–6	0.4027	0.4416	0.4116	9.66	2.21
Case 5	4–12	6–10	0.1091	0.1689	0.1126	67.10	3.20
		4–6	0.4027	0.6208	0.4100	54.17	1.81
		12–13	0.1091	0.1689	0.1126	14.19	3.20
		2–6	0.3802	0.4201	0.3993	10.47	5.02

TABLE 5.7

Comparison of Real Losses with Original Schedule and Reschedule of GENCOs during Contingency

	P_L in p.u. with		% Reduction in Active Line Loss with
Cases	Original Contribution of GENCOs	Recontribution of GENCOs	Recontribution of GENCOs
Case 1	0.0798	0.0733	8.14
Case 2	0.1427	0.1236	13.38
Case 3	0.0891	0.0775	13.01
Case 4	0.0809	0.0742	8.28
Case 5	0.0813	0.0719	11.48

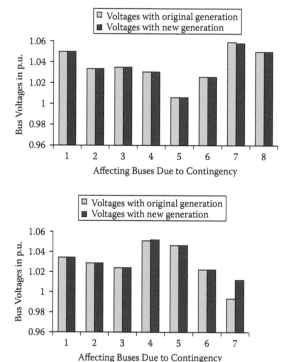

FIGURE 5.9

Comparison of voltage during contingency with original and new generation for cases 1 and 2.

Hence, a multi-objective methodology of DE- and PSO-based optimization has been considered that, along with the congestion and voltage security constraints, takes boundary conditions of generation cost and hence electricity price into consideration. A controlled load curtailment technique has been adopted to implement the optimization technique in deregulated

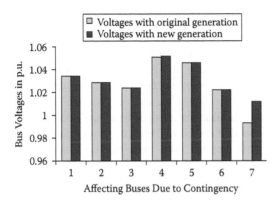

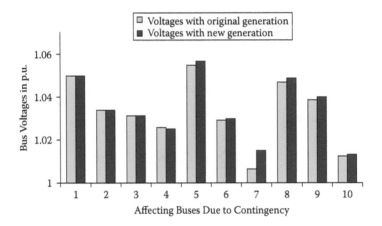

FIGURE 5.10
Comparison of voltage during contingency with original and new generation for cases 3 and 4.

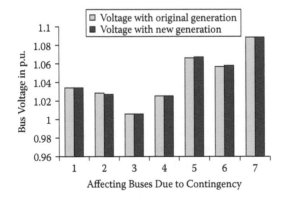

FIGURE 5.11
Comparison of voltage during contingency with original and new generation for case 5.

power systems [310]. As load curtailment can reduce the reliability of power supply, this optimization technique assures the least variation of system security by keeping curtailment to a minimum possible limit predicted by both DE and PSO algorithms. Hence, the participants can avoid the variation of generation cost, and thus price electricity with a minimum possible disturbance to system reliability.

Minimum operating cost and minimum shift from the optimum operation have been used as the objective functions for the proposed multi-objective function. Considering these criteria, the problem can be formulated as follows.

Minimize

$$OBF = \sum_{i \in NG} C_i(P_{Gi}) + \sum_{i \in NG} \Delta C_i \text{ US\$/h} \tag{5.29}$$

where

$$\Delta C_i = C_{0i} - C_i^j ; C_i(P_{Gi}) = (\alpha_i P_{Gi}^2 + \beta_i P_{Gi} + \gamma_i) \tag{5.30}$$

subject to the constraints

$$P_{Gi} - P_{Di} = V_i \sum_{j \in N_i} V_j \left(G_{ij} \cos\theta_{ij} + B_{ij} \sin\theta_{ij} \right) \tag{5.31}$$

$$Q_{Gi} - Q_{Di} = V_i \sum_{j \in N_i} V_j \left(G_{ij} \sin\theta_{ij} - B_{ij} \cos\theta_{ij} \right), i = 1,2,..., \text{ NB} \tag{5.32}$$

$$P_{Gk} = P_{Gk}^C \quad k = 1, 2, ..., \text{NG} \tag{5.33}$$

$$P_{Dj} = P_{Dj}^C \quad j = 1, 2, ..., \text{ND} \tag{5.34}$$

$$P_{Gi}^{\min} \le P_{Gi} \le P_{Gi}^{\max} \tag{5.35}$$

Security constraint: $V_i^{\min} \le V_i \le V_i^{\max}$ and $P_{ij\min} \le P_{ij} \le P_{ij\max}$ \hfill (5.36)

where C_{0i} and C_i^j are the generation costs of each generator for the base transaction and contingent state transaction, respectively.

To calculate the generator rescheduling, the developed algorithm has been applied in the IEEE 30 bus test system for the worst possible contingencies. The harmful contingencies can be found by proper weak bus selection (by using (3.41)). From that selection, it has been found that bus 30 (second weakest bus) has connections to both buses 27 and 29. Hence, line 29–30 has been tripped. The tripping procedure continued for several other cases.

The solution methodology, adopted under the prevalence of these cases, is shown in Figure 5.12.

According to the flowchart of the solution methodology, as expected, with DE optimization, in this contingent condition the generation schedule changes and the generation cost increases (Table 5.8). The PSO-based

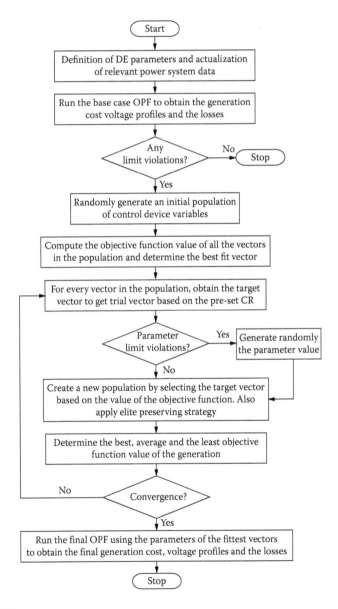

FIGURE 5.12
Flowchart of the solution methodology.

TABLE 5.8

Rescheduled Dispatch of Generators for Different Contingent States

	Line Trip between the Buses	Rescheduled Dispatch (MW)						Optimized Generation Cost (US$/h)
		Generator Bus No.						
		1	2	5	8	11	13	
DE-based OPF	29–30	176.78	48.84	21.47	21.68	12.10	12.00	802.2851
	29–27	176.90	48.87	21.48	21.75	12.12	12.00	803.1148
	12–14	176.79	48.84	21.48	21.72	12.15	12.00	802.6358
	12–16	176.59	48.81	21.47	21.74	12.26	12.00	802.2926
	16–17	176.64	48.81	21.47	21.68	12.17	12.00	801.9346
PSO-based OPF	29–30	176.81	48.79	21.48	21.71	12.10	12.00	802.2852
	29–27	176.64	48.93	21.49	21.89	12.15	12.00	803.1153
	12–14	176.78	48.86	21.53	21.65	12.08	12.00	802.6362
	12–16	176.84	48.74	21.50	21.60	12.20	12.00	802.2932
	16–17	176.66	48.86	21.47	21.59	12.21	12.00	801.9348

optimization also confirms the above fact. This increase in generation cost violates the previous agreement between market participants, and ISO has to place new proposals to the participants, which may be unacceptable for any one of them. Again, with contingency, not only the GENCOs are suffering due to higher operating cost, but TRANSCOs and DISCOs are also suffering due to high congestion management cost.

To limit the generation cost according to the developed algorithm, the DE- and PSO-based OPF searches for an optimum schedule until the operating cost goes below the precontingent value. For each of these simulations, load has been curtailed according to willingness to pay ranking, and the technique adopted is just like pool curtailment, where the main plan is to minimize the deviation of the transactions from the desired values. In deregulated environments, there are basically two kinds of contracts: firm and non-firm. Non-firm contracts have less willingness to pay and hence are subject to curtailment. In this market environment, reliability will appear as a commodity to be sold at the price of spot price volatility. The results are shown in Table 5.9.

As observable in that table, the price volatility can be indirectly controlled by the optimal load curtailment technique. The market price of electricity can thus be withheld by efficient load curtailment even in the worst conditions of contingency.

The generation cost more or less remains the same with DE and PSO algorithms. In some cases PSO-based results offer more reliability with less curtailment of load but higher generation cost. The ISO can choose any one of the solutions depending upon the market conditions and reliability issues. But with DE- or PSO-based results, as the generation cost remains the same, there can be the least variation of spot price.

TABLE 5.9

Rescheduled Dispatch of Generators for Different Contingent States with Load Curtailment to Limit Generation Cost

		Line Trip	Load Curtailed at Bus	% Load Curtailed	Rescheduled Dispatch (MW)						Optimized Generation Cost (US$/h)
					Generator Bus Number						
					1	2	5	8	11	13	
DE-based developed algorithm	1	29–30	29	9.7500	176.67	48.81	21.46	21.61	12.08	12.0	801.4273
	2		30	2.5000	176.23	48.81	21.46	21.59	12.07	12.0	801.2529
	3	12–14	12	2.5000	176.65	48.81	21.47	21.65	12.12	12.0	801.6405
	4		14	4.9375	176.63	48.80	21.46	21.63	12.12	12.0	801.4873
	5	14–15	14	2.5000	176.64	48.81	21.46	21.60	12.08	12.0	801.3042
	6		15	2.5000	176.62	48.80	21.46	21.59	12.07	12.0	801.1175
	7	12–16	12	2.5000	176.45	48.77	21.46	21.66	12.24	12.0	801.3007
	8		16	4.9375	176.50	48.78	21.47	21.69	12.24	12.0	801.6593
PSO-based developed algorithm	1	29–30	29	7.3141	176.67	48.83	21.44	21.69	12.09	12.0	801.6416
	2		30	2.5000	176.65	48.85	21.46	21.53	12.11	12.0	801.2530
	3	12–14	12	2.5000	176.59	48.88	21.48	21.54	12.23	12.0	801.6409
	4		14	4.9375	176.71	48.82	21.48	21.61	12.05	12.0	801.4874
	5	14–15	14	2.5000	176.74	48.78	21.52	21.50	12.08	12.0	801.3045
	6		15	2.5000	176.48	48.82	21.49	21.56	12.17	12.0	801.1179
	7	12–16	12	2.5000	176.57	48.70	21.45	21.56	12.24	12.0	801.3008
	8		16	4.9375	176.46	48.85	21.50	21.59	12.30	12.0	801.6596

As the minimum possible load curtailment is promoted, the method aims to maintain maximum reliability of the power system in deregulated environments. However, the ISO can also offer a schedule without load curtailment but with higher price during contingency (Table 5.8). In that case, all the participants have to agree upon this new transaction and compromise about the price hike. The results, shown in Table 5.9, are quite satisfactory, as it not only bounds the generation cost, but also relieves the lines from overloading, and hence manages congestion. The overload relief of these lines is depicted in Table 5.10.

The bus voltage profiles for the base case, contingent case and contingent case with controlled load curtailment are shown in Figure 5.13. The precontingent (base case) voltage profile has been subjected to a serious decline after the artificial implication of worst possible contingencies. This sag of voltage can be revived by minimum alteration of the generation schedule with an optimal load curtailment. The transfer capability relies heavily on voltage profile and can be improved in the contingent state by effective utilization of this methodology.

It can be inferred that as the algorithm not only observes the economic aspect but also tests the feasibility of rescheduling by running load flow for each of the cases, the results are always within stability limits and operating loss is minimum. The above methodology, being adopted by ISO, can create a fair competition between the market participants; as with load curtailment, the revenue is lost and hence the participants should improve their security levels so that the contingencies may not arise at all in the system.

TABLE 5.10

Overload Relief of Congested Lines with Rescheduled Dispatch

			Real Power Flow (MW)			
Line Trip	Two Most Congested Lines	Without Line Trip	Line Trip but Same Generation Case I	Line Trip but with New Generation and Load Curtailment Case II	% Overload Factor (Case I)	Improvement in Percentage Overload Factor (Case II)
29–30	27–30	7.1063	11.4982	9.8075	61.8029	38.01
	28–27	14.2988	14.7609	14.0079	3.2317	–2.03
27–30	29–30	3.7071	10.9500	5.9520	195.379	60.55
	28–27	14.2988	14.8632	14.0002	3.9472	–2.08
29–27	28–27	14.2988	14.8632	14.2307	3.9472	–0.47
	27–30	7.1063	13.7018	11.6543	92.812	63.99
12–14	12–15	17.3815	23.6686	19.9887	36.1712	14.99
	12–16	5.8184	7.5406	5.5002	29.5992	–5.46
14–15	14–12	5.6612	6.2456	5.8654	10.3229	3.60
	12–15	17.3817	18.5233	18.0038	6.5678	3.57

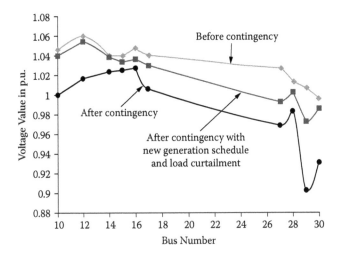

FIGURE 5.13
Bus voltage profile for precontingent, contingent and load curtailment conditions.

Actually, contingency is an inevitable fact, and along with preventive measures, some corrective algorithms are to be developed for the maintenance of both reliability and security of the system. The following section describes one more methodology to mitigate line congestion without any high investment like flexible AC transmission system (FACTS) and high-voltage direct current (HVDC) or reliability degrading procedure like load curtailment.

5.5.3 Mitigation of Line Congestion and Cost Optimization Using Multi-Objective PSO

This section presents another effective method of congestion management. As discussed earlier, congestions or overloads in transmission networks are mitigated by generation rescheduling or load curtailment. Here, the two conflicting objectives, (1) mitigation of overload and (2) minimization of cost of operation, have been optimized to provide pareto-optimal solutions. A new multi-objective particle swarm optimization (MOPSO) method is used to solve this complex non-linear optimization problem [309]. This congestion-constrained cost optimization algorithm is capable of limiting line congestion with a minimum management charge without any load curtailment, and it also provides better operating conditions with respect to voltage profile, total line loss and security for the system during contingency. For contingency selection and ranking, a security sensitivity index (SSI) has also been developed. The main advantage of this algorithm is that it can also confine the level of congestion at any preferred value according to the affordable congestion management cost decided by market participants in deregulated environments. Though the algorithm (Figure 5.14) has been

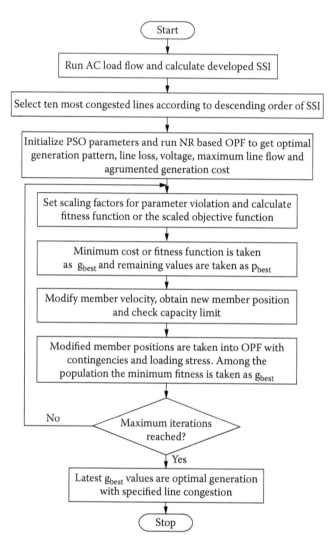

FIGURE 5.14
Flowchart of the developed solution methodology.

shown to be tested on the IEEE 30 bus test system, it has been found to be effective with systems with a higher number of buses.

Power system security can be defined as its ability to maintain a normal state even during contingency. The security level can be judged by a developed index named the security sensitivity index (SSI). Mathematically, it can be expressed as follows:

$$SSI = \frac{P_{ij}^0 - P_{ij}^j}{P_i^0 - P_i^j} \tag{5.37}$$

where P_{ij}^0 and P_{ij}^j are the line flows in MW between bus i and bus j before and after tripping, and P_i^0 and P_i^j are total power injected to bus i before and after tripping. It is quite evident that the higher the value of this index, the lower is the level of security of the line.

The methodology has been primarily used with conventional cost optimization, and then it has been applied with the developed congestion-constrained cost optimization problem using PSO. The equality, inequality and security constraints, however, remain the same for the conventional and developed algorithms. In this algorithm, the objective function has been modified by a scaling factor, which appears only during line flow constraint violates. As the optimization process tries to minimize the objective function, the scaling factor associated with maximum line congestion directs it to produce a new generation schedule, which offers the specified line flow.

The objective function for the developed optimization is

$$\text{Minimize } F = \sum_{i=1}^{NG} C_i + sf \times P_{ij\max} \text{ \$/h} \tag{5.38}$$

The constraints are as follows:

Equality or power balance constraints:

$$\sum_{i=1}^{NG}\sum_{g\in G}\left(P_{gi}^0 + \Delta P_{gi}^j\right) - \left(P_l^0 - LCP^j\right) = \sum_{k=1}^{N}\left|V_i^j\right|\left|V_k^j\right|\left|Y_{ik}^j\right|\cos\left(\theta_i^j - \theta_k^j - \delta_{ik}^j\right) \tag{5.39}$$

$$\sum_{i=1}^{NG}\sum_{g\in G}\left(Q_{gi}^0 + \Delta Q_{gi}^j\right) - \left(Q_{li}^0 - LCQ^j\right) = \sum_{k=1}^{N}\left|V_i^j\right|\left|V_k^j\right|\left|Y_{ik}^j\right|\sin\left(\theta_i^j - \theta_k^j - \delta_{ik}^j\right) \tag{5.40}$$

Inequality or generator output constraints:

$$P_{Gi}^{\min} \leq P_{Gi}^0 \leq P_{Gi}^{\max} \tag{5.41}$$

$$Q_{Gi}^{\min} \leq Q_{Gi}^0 \leq Q_{Gi}^{\max} \tag{5.42}$$

$$\Delta P_{Gi}^{\min} \leq \Delta P_{Gi}^j \leq \Delta P_{Gi}^{\max} \tag{5.43}$$

$$\Delta Q_{Gi}^{\min} \leq \Delta Q_{Gi}^j \leq \Delta Q_{Gi}^{\max} \tag{5.44}$$

Voltage constraint and transmission constraints are the same as described in (5.36).

Load curtailment limits:

$$0 \leq LCP^j \leq P_l^{\max} \tag{5.45}$$

$$Q_{li}^{\min} \leq LCQ_i^j \leq Q_{li}^{\max} \tag{5.46}$$

The developed methodology has been depicted in the flowchart (Figure 5.14). In the first step AC load flow has been carried out to identify the 10 most congested lines in the network by the developed SSI. After the evaluation of constraints violation, a scaling factor is multiplied with that to obtain the objective function as described earlier. Now the PSO-based parallel search method, instead of looking for a generation pattern for minimum generation cost, quests for a solution, which keeps minimum cost with the preset maximum loading limit. Being a stochastic optimization technique, the non-linearity of the constraints cannot restrain PSO to reach the global optima. Hence, for this optimization PSO has been given utter importance over the other prevalent optimization methods, like Lagrange's multiplier or Poyntriangen's principle. After initialization of random solution vectors, the PSO-based search method in each iteration evaluates the fitness function to get p_{best} and g_{best} values. The termination of iteration is obtained from the predefined accuracy criteria or number of iterations.

For congestion management, the most vulnerable lines of the IEEE 30 bus test system with respect to congestion need to be selected, as they are the possible weak links of the system. Figure 5.15 represents 17 lines with their SSIs in descending order.

The algorithm has been applied to the system along with the conventional optimization method for different contingencies. During contingency, when the line flow exceeds a certain limit, a scaling factor is added with the objective function to be minimized (5.38). This in effect tries to minimize congestion along with the generation cost. Table 5.11 depicts the comparison

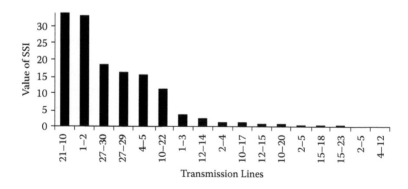

FIGURE 5.15
Selection of vulnerable lines by developed SSI in the IEEE 30 bus system.

TABLE 5.11

Comparison of Line Flows

Tripped Cases	Line Connecting the Buses	Maximum Line Flow (MW) Calculation Using PSO-Based Optimization	
		Without Congestion Management	With Developed Congestion-Constrained Cost Optimization Methodology
1	10–21	118.632	71.9999
2	1–2	151.2315	71.9985
3	27–30	118.6838	71.9990
4	27–29	118.6488	71.9999
5	4–6	129.6440	71.9915
6	10–22	118.5621	71.9998
7	1–3	169.5659	71.9999
8	12–14	118.6401	71.9978
9	2–4	105.2676	71.9973
10	10–17	118.4848	71.9983

of the line flow computed by the conventional and the developed method with scaling factor.

Though here the line flow limit has been taken as a forecasted value of 72 MW (SIL/2), the algorithm can limit the line flow at any specified value. In a deregulated environment, the ISO can use this algorithm to reschedule the GENCOs for the required level of congestion management.

Though this developed method effectively reduces line flow without any load curtailment, the total generation cost increases due to the change in individual contribution of the generators. Hence, the new schedule of generation can be imposed in the immediate hour or on an hourly basis. Table 5.12 depicts the comparison of generation costs obtained from conventional OPF with developed methodology with a scaling factor. The rescheduling of generation can bring about congestion relief, but the generation contribution cost increases. This cost, however, can be recovered from the market participants as security charge.

Table 5.13 shows the variation of congestion management cost for limiting the line congestion at 50% of the surge impedance loading (SIL) level. The same congestion management may be achieved by load curtailment according to willingness to pay, but only with a high management cost of losing load. In this case, value of lost load (VOLL) is the congestion management cost.

Figure 5.16 depicts congestion management and corresponding cost for single-line (1 ~ 3) contingency. The lower limit of line flow can ensure excess power flow handling capacity in stressed conditions as well as an escalation in new transactions catering capacity. But again, it is quite evident that the higher the level of allowed congestion, the lower is its management cost.

TABLE 5.12

Comparison of Generation Costs

Tripped Cases	Line Number	Total Generation Cost ($/h) Using PSO	
		Without Congestion Management	With Developed Congestion-Constrained Cost Optimization
1	10–21	803.0751	836.3210
2	1–2	839.2833	895.2044
3	27–30	903.5036	936.8287
4	27–29	803.1002	836.3659
5	4–6	806.6264	848.1328
6	10–22	802.0998	835.2446
7	1–3	815.1886	891.3969
8	12–14	802.6182	835.8944
9	2–4	804.0127	822.7905
10	10–17	802.3837	835.4275

TABLE 5.13

Calculation of Congestion Management Cost

Tripped Cases	Tripped Lines	Congestion Management Cost ($/h) Using Developed Algorithm	Value of Lost Load (VOLL) or Congestion Management Cost ($/h) with Load Curtailment
1	10–21	33.2459	198.09
2	1–2	55.9211	225.39
3	27–30	33.3251	185.70
4	27–29	33.2657	183.67
5	4–6	41.5064	184.01
6	10–22	33.1448	183.61
7	1–3	76.2083	271.95
8	12–14	33.2762	183.65
9	2–4	18.7778	147.73
10	10–17	33.0438	189.00

Another important objective of contingency management other than congestion management in real-time system operation is to maintain the voltage profile and minimum possible line loss. Figure 5.17 shows the comparison of bus voltage profiles between the above-mentioned two methods. The voltage profile with the developed method is better than the conventional cost optimization method. The method also offers an advantage of reduced line loss. As shown in Figure 5.18, the line loss has reduced considerably with respect to the conventional method. Hence, before considering the congestion management cost, ISO should consider the long-term effects of reduced line loss.

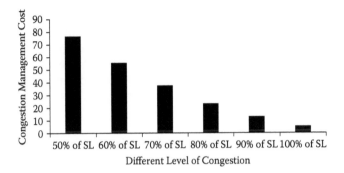

FIGURE 5.16
Congestion management cost for different levels of congestion.

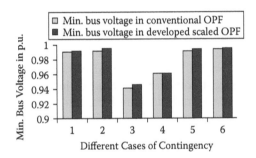

FIGURE 5.17
Comparison of voltage profiles.

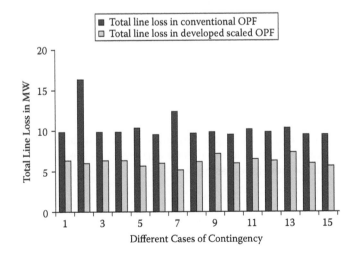

FIGURE 5.18
Comparison of line losses.

Though this method recovers the system from contingency it introduces an additional congestion management cost that may not be recovered from the participants by ISO. Hence, a multi-objective congestion-constrained cost and curtailment optimization algorithm needs to be formulated along with some factors for the estimation of level of contingency during an eventuality.

5.5.4 Swarm Intelligence-Based Cost Optimization for Contingency Surveillance

In this section, a new multi-objective optimization model has been used to minimize congestion cost, load curtailment and generation cost simultaneously to restore the operating point of the system under contingency. The solution algorithms of this method are based on the particle swarm optimization (PSO), in which load curtailment and generation cost have been optimized without breaching line flow constraints for congestion management. The significance of this method has been presented by a comparative study with the conventional cost optimization method in terms of operating cost considering value of lost load (VOLL) and two more newly developed analytical indices, namely value of congestion cost (VOCC) and value of excess loss (VOEL), in contingent states of the power system [313]. The generation cost of power is not the sole determining factor of the operating cost of the power system. The congestion management cost, the value of lost load and the value of excess loss need to be incorporated in the operating cost of a system. The system operator has to effect rescheduling only after considering the rise in the operating cost of a system rather than only generation cost. These analytical indices have an important role during operating cost of power generation calculation. These analytical indices and their adopted formulations are as cited below.

5.5.4.1 Development of Value of Lost Load (VOLL)

VOLL is the estimated amount that the consumer receiving electricity with firm contracts would be willing to pay to avoid a disruption in its electricity service. The value of these losses can be expressed as the customer damage function. It can be developed as below:

$$VOLL = \frac{\sum_{i=1}^{NG} C_i^j}{\sum_{i=1}^{NG} P_{Gi}} \left(P_i^0 - LCP^j \right) \tag{5.47}$$

5.5.4.2 Development of Value of Congestion Cost (VOCC)

Due to increase in demand unexpected congestion bottlenecks may be created in the system. To reduce congestion the system operator has to charge for excess flow of power through the connecting lines. In a particular zone, receiving power not only pays the generation cost but also has to pay the congestion charge (if any) associated with the transmission. The value of congestion cost (VOCC) can be formulated as

$$VOCC = \left[\frac{\sum\limits_{i=1}^{NG} C_i^j + VOLL}{\left(P_i^0 - LCP^j \right) + \left(\sum\limits_{i=1}^{NG} P_{Gi}^0 - P_i^0 \right)} - \frac{\sum\limits_{i=1}^{NG} C_i^j}{P_i^0} \right] \times \left[P_{ij}^j - P_{ij}^0 \right] \qquad (5.48)$$

5.5.4.3 Development of Value of Excess Loss (VOEL)

The line loss is an additional important factor of the operating cost. GENCOs have to generate this excess power to meet the demand in the system. If a system operates at a higher value of line loss than its optimal value, then the excess loss has to be incorporated in the operating cost. VOEL can be developed as

$$VOEL = \frac{\sum\limits_{i=1}^{NG} C_i^j}{\sum\limits_{i=1}^{NG} P_{Gi}} \left\{ \left(\sum\limits_{i=1}^{NG} P_{Gi}^0 - P_i^0 \right) - \left(\sum\limits_{i=1}^{NG} P_{Gi}^j - \left(P_i^0 - LCP^j \right) \right) \right\} \qquad (5.49)$$

The objective function for the developed multi-objective congestion-constrained cost and load curtailment optimization is as follows with all other equality and inequality constraints, which have already been described earlier in this chapter.

Minimize

$$F = \sum\limits_{i=1}^{N_G} C_i^j \left(P_{gi}^0 + \Delta P_{gi}^j \right) + LC_i^j \left(P_i^0 - LCP^j \right) + CC_i^j \left(P_{ij}^j - P_{ij}^0 \right) \qquad (5.50)$$

where CC_i^j is the cost of unavoidable congestion and LC_i^j is the cost of curtailed load.

To establish the effectiveness of this multi-objective congestion-constrained cost and load curtailment optimization, two more conventional optimization techniques have been considered:

Method 1: Cost optimization without load curtailment

Method 2: Cost optimization with load curtailment

Table 5.14 shows the comparison of developed optimization (5.49) result with the other two conventional optimizations applicable for a deregulated power market. The results of the first method, which is the conventional cost optimization, show that curtailment can be avoided by rescheduling, resulting in increased generation cost, line flow and transmission loss. In quite a few cases the line flow exceeds even the SIL limit, which in the future can

TABLE 5.14

Optimization Results for the Conventional and Developed Methods

Methods	Line Trip	Gen Cost (US$/h)	Max Line Flow (MW)	% Load Curtailment	Trans Loss (MW)
Cost optimization without load curtailment	2–4	804.01	105.25	0.0	9.881
	6–7	806.70	125.11	0.0	10.81
	2–6	805.86	104.36	0.0	10.26
	4–6	806.72	129.31	0.0	10.35
	3–4	814.41	167.43	0.0	12.14
	1–3	815.18	169.52	0.0	12.31
	2–5	836.44	103.57	0.0	18.28
	1–2	839.28	151.02	0.0	16.16
	2–4, 1–3	825.78	163.97	0.0	14.31
	2–4, 1–3, 2–6	873.55	167.27	0.0	22.19
	2–4, 1–3, 2–6, 4–6	873.23	162.54	0.0	20.72
Cost optimization with load curtailment	2–4	800.55	104.81	0.33	9.8121
	6–7	801.03	125.11	0.5	10.335
	2–6	800.69	103.84	0.5	10.179
	4–6	801.52	128.62	0.5	10.264
	3–4	800.63	165.61	1.31	11.891
	1–3	799.88	167.39	1.63	12.007
	2–5	799.77	97.15	2.80	12.912
	1–2	799.31	147.59	3.55	15.189
	2–4, 1–3	800.21	160.85	2.43	13.86
	2–4, 1–3, 2–6	800.03	148.91	6.10	18.30
	2–4, 1–3, 2–6, 4–6	799.41	150.95	6.77	19.16
Developed multi-objective congestion-constrained cost and load curtailment optimization	2–4	804.01	105.25	0.0	9.88
	6–7	739.40	117.34	6.31	9.07
	2–6	805.86	104.36	0.0	10.25
	4–6	690.47	114.83	11.42	8.49
	3–4	519.95	117.57	30.27	7.39
	1–3	515.63	117.62	30.82	7.43
	2–5	836.44	103.57	0.0	18.32
	1–2	545.18	117.68	28.23	9.35
	2–4, 1–3	512.71	117.90	32.16	10.25
	2–4, 1–3, 2–6	596.45	117.17	27.30	16.91
	2–4, 1–3, 2–6, 4–6	599.96	115.25	27.30	17.25

initiate a cascading failure of the system. Moreover, congestion charges can rise to a high value along with transmission loss charge. This method thus achieves high reliability in terms of load sustainability at the cost of a wide range of price volatility.

The second method is similar to conventional cost optimization; the only difference is that a controlled load curtailment is affected if the optimized cost goes above a precontingent value. This method is capable of sustaining the generation cost for consumer welfare, but as shown, the degree of congestion is well beyond an affordable limit, and also the curtailment has not been optimized with respect to optimized generation cost. Though this method attempts to limit the price volatility by limiting generation cost, congestion cost and the value of excess loss can increase the price volatility. The projected multi-objective optimization method takes congestion as the deterministic criterion of load curtailment, and the same has been optimized with generation cost optimization. Hence, the method minimizes load curtailment, congestion and generation cost simultaneously.

The rise in operating cost of electricity is a combined effect of congestion charge (VOCC), excess power loss (VOEL) and value of lost load (VOLL). With respect to the conventional cost optimization method, these charges have been calculated and the total expected operating cost has been compared with that of method 2 and the developed congestion-constrained load curtailment optimization (Table 5.15). The operating cost is the sum of generation cost and VOLL, VOCC and VOEL.

TABLE 5.15

Comparison of Operating Cost Considering VOCC, VOLL and VOEL

Line Trip	Cost Optimization without Load Curtailment (Method 1)	Cost Optimization with Load Curtailment (Method 2)	Developed Multi-Objective Congestion-Constrained Cost and Load Curtailment Optimization
	Total Operating Cost (US$/h)	Total Operating Cost (US$/h)	Total Operating Cost (US$/h)
2–4	804.01	805.86	804.01
6–7	812.46	807.61	787.56
2–6	805.87	807.27	805.86
4–6	815.39	808.28	776.60
3–4	869.47	815.06	737.52
1–3	873.57	815.21	737.08
2–5	836.34	822.48	836.44
1–2	882.49	831.41	750.20
2–4, 1–3	871.52	868.42	743.50
2–4, 1–3, 2–6	914.76	877.67	803.48
2–4, 1–3, 2–6, 4–6	804.01	887.34	807.89

The multi-objective optimization offers comparatively less operating cost with respect to the other two methods. Method 1 has been charged with congestion cost and value of excess loss as it fails to manage congestion and offers higher cost. This method can be adopted by ISO and market participants, but only at the cost of congestion and line loss. Method 2, however, offers less cost in comparison with method 1 with an additional VOLL as it tries to sustain the generation cost at a fixed value by load curtailment. This method, though, offers less volatility of operating cost, but reliability is affected and the participants who are less willing to pay can agree with this schedule. In method 2, VOLL is less than the developed method, as it does not have any congestion constraint. Thus, this method produces less generation cost, and hence participants who are less willing to pay can adopt it. Figures 5.19 to 5.21 show the variation of VOLL, VOCC and VOEL charges of the conventional and developed optimization methods.

Thus, the operating cost of the power is the algebraic summation of generation cost and these three indices. Figures 5.22 to 5.24 depict the money and information flow among GENCOs, TRANSCOs and DISCOs for the three different methods.

Thus, this PSO-based algorithm is helpful for congestion management in a contingent system with optimum load curtailment and generation cost. On violation of a stipulated line flow, a controlled load curtailment has been adopted to minimize congestion, and the objective function has been represented as a multi-objective minimization problem in a PSO environment to minimize curtailment as well as generation cost.

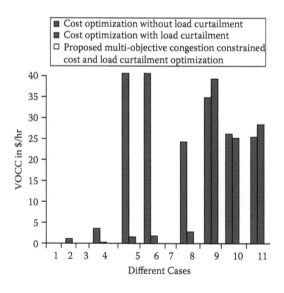

FIGURE 5.19
Comparison of value of congestion cost.

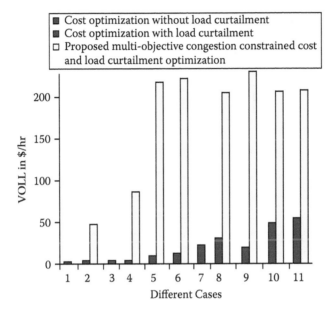

FIGURE 5.20
Comparison of value of lost load.

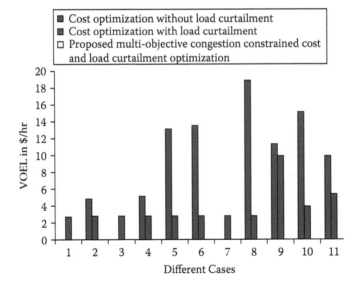

FIGURE 5.21
Comparison of value of excess loss.

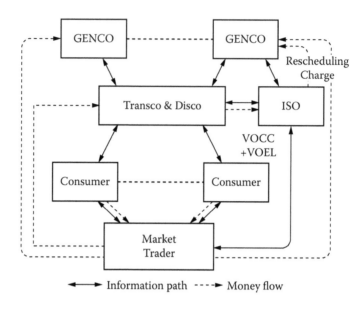

FIGURE 5.22
Information and money flow according to method 1.

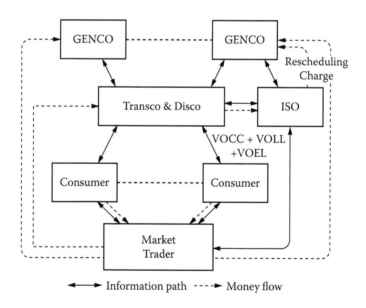

FIGURE 5.23
Information and money flow according to method 2.

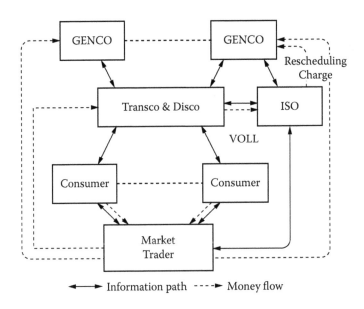

FIGURE 5.24
Information and money flow according to method 3.

5.6 Summary

The increasing demand for an OPF tool for assessing state and recommended control actions for off-line and on-line regulation of operating conditions has reached an optimum level. Since the first OPF paper was presented in the 1960s, the increase in need for OPF to solve the problems of today's deregulated industry and the unsolved problems in the vertically integrated system have caused the working group on operating economics to evaluate the capabilities of existing OPF. OPF has enjoyed the renewed interest in a variety of formulations through the use of advanced optimization techniques such as GA, PSO and DE. In spite of all the investment and work to date on OPF, several questions have remained unanswered and have become a challenge to power engineers over the last decade. The emergence of OPF was only categorized as a tool of producing optimal generation patterns for minimal cost. This idea is getting modified as OPF is no longer related only to cost optimization; rather, the objective has included more operational constraints like line congestion, line loss and price volatility. In this respect the multi-dimensional OPF presented in this chapter is evolutionary and has been effectively used to enrich the performance of the network in normal as well as contingent states of operations. The methodologies employed can be applicable to include more constraints as they emerge day by day. Hence, the future prospects of these methodologies will lie in effective utilization and implementation.

Annotating Outline

- Due to the increase in demand and constraints, new multi-objective methodologies and algorithms are to be harnessed in the field of power engineering. The conventional optimization algorithms like generation cost and loss minimization are one-dimensional and do not incorporate limit violation. Issues like congestion management, price volatility and operational cost minimization are to be given equal importance in modern-day power markets.

- The multi-objective line loss optimization algorithm developed effectively utilizes GA to search for an optimum solution in a deregulated environment (Table 5.3).

- It has been demonstrated that GA can be effectively utilized to regulate the generator contribution for the management of line congestion in a deregulated system (Table 5.6).

- An optimal load curtailment-based price volatility minimization has also been depicted for reschedule of generation pattern without any limit violation using DE and PSO (Table 5.10).

- It has been cited that load curtailment can be avoided and simultaneous congestion management can be achieved by redispatching generation without any compensating devices (Table 5.12).

- For the identification of congested lines the overloading index (OI) and security sensitivity index (SSI) have been developed for normal and contingent states, respectively.

- The performance evaluation of OPFs is imperative from the point of view of selection for a particular power network. VOLL, VOCC and VOEL, in this chapter, effectively analyze OPFs and develop a novel multi-dimensional OPF considering all the objectives and constraints of the present-day power network.

6

Application of Stochastic Optimization Techniques in the Smart Grid

The term *smart grid* refers to modernization of the power network so that it monitors, optimizes and can take preventive action in the operation of its interconnected elements – from the central generation through the high-voltage transmission network and the distribution system, to the industrial and domestic user. Ongoing development of smart metering has made demand response an indispensable part of real-time operation and control of modern grids. The optimization not only looks for a generation schedule, but also demands consumer welfare in terms of distributed load curtailment, less penalty and formidable load schedule. The insight of this chapter develops a curtailment index (CI) to trace the eligibility of the participants to retain their position in the market and an acceleration factor (AF) to hasten the optimization with the required level of accuracy. Instead of penalising for concentrated increase of demand, the described optimization approach effectively segregates the participants to impose a distributed load curtailment to reduce the volatility of price of electricity in the market.

6.1 Introduction

With the promotion of world economy modernization, the price of oil has been kept on an upward trend. What is also noticeable is the shortage of energy supply around the world, the increasing pressures on resources, environment and the enormous power losses in energy delivery due to the low efficiency of the current power grid. What is more, owing to the growing electricity demands and the users' increasing requirements for reliability and quality, the power industry is now facing unprecedented challenges and opportunities. Hence, the existing power system is to be transformed to become more environment friendly, economic, safe, reliable and flexible.

The emergence of advanced metre infrastructure and its extensive use with Internet facility envisage the acceleration of this allotropy [302]. Since the 1990s, with the increasing use of distributed generation power, more demands and requirements have been proposed for power grid intensity [303, 304]. To find out an optimal solution for these problems, power companies should accept the idea of new technology adoption, potential mining of the existing power system and improvement of its application and utilization. Consensus has been reached by experts and scholars from different countries that future power grids must be able to meet various requirements of energy generating units and the demands of highly market-oriented power transactions so that the needs of the self-selection from customers can be satisfied individually. All of these will become the future development direction of the smart grid. Actually, the smart grid integrates and enhances other essential elements, including traditional upgrades and new grid technologies with renewable generation, storage, increased consumer participation, communications and computational ability. Basically, the smart grid will be planned to ensure high levels of security, quality, reliability and availability of electric power. It also improves economic productivity and quality of life, and minimizes environmental impact while maximizing safety and social welfare. Characterized by a two-way flow of electricity and information between utilities and consumers, the smart grid will deliver real-time information and enable the near-instantaneous balance of supply and demand.

In this chapter, a few cost-effective optimal power flow (OPF) models based on stochastic techniques have been described for maximizing the utilization of available resources with the introduction of smart metering technologies, as is expected in the smart grid arena.

6.2 Smart Grid and Its Objectives

The smart grid envisioned taking advantage of all available modern technology in transforming the current grid to one that functions more intelligently to facilitate better situational awareness and operator assistance, autonomous control action to enhance the reliability by increasing resiliency against component failure, efficiency enhancement by maximizing asset utilization, integration of renewable energy sources, real-time communication between consumers and utility and improved market efficiency through innovative solutions and higher quality of service.

The sub-sections below are dedicated to highlighting the proposed concept of the smart grid and the requirement of amalgamation of these technologies with optimization techniques.

6.2.1 Concept of the Smart Grid

The smart grid is a gradual development process accompanied with technology innovation, demands of energy saving and management needs. People will have their own understanding of the smart grid, no matter if they are

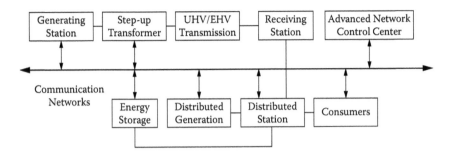

FIGURE 6.1
Conceptual diagram of the smart grid.

facility suppliers, IT companies, consulting firms, public power companies or power generation companies. From the earlier smart intelligence metering to electrical intelligence, from transmission and distribution automation to a whole intelligent process, the concept of the smart power grid has been enriched substantially. In 2006, U.S. IBM presented a 'smart grid' solution. This is a relatively complete concept for the current smart grid, which indicates its official birth. As shown in Figure 6.1, a smart grid is basically overlaying the physical power system with an information system that links a variety of equipment and assets together with sensors to form a customer service platform. It allows the utility and consumers to constantly monitor and adjust electricity use.

The management of operation will be more intelligent and scientific based on the dynamic analysis of needs from both the user side and the demand side, which can increase capital investment efficiency due to tighter design limits and optimized use of grid assets. In comparison with the traditional grid, the smart grid includes integrated communication systems, advanced sensing, metering, measurement infrastructure, complete decision support and human interfaces.

6.2.2 Elementary Objectives of the Smart Grid and Demand Response

The momentum for the smart grid vision has increased recently due to policy and regulatory initiatives. Numerous and diverse stakeholders are striving to realize the above smart grid goals by advancing and deploying various technologies. These efforts can be categorized into the following trends:

- Reliability
- Renewable resources
- Demand response
- Electric storage
- Electric transportation

The above trends are also recognized as priority functional areas in the Federal Energy Regulatory Commission (FERC) 'Smart Grid Policy' statement [305]. Among these trends, system reliability has always been a major focus area for the design and operation of modern grids. The other trends involve distinct smart grid resource types with diverse impacts on reliability. Renewable resources, while supplementing the generation capability of the grid and addressing some environmental concerns, aggravate the reliability due to their volatility. Demand response and electric storage resources are necessary for addressing the economics of the grid and are perceived to support grid reliability through mitigating peak demand and load variability. Balancing the diversity of the characteristics of these resource types presents challenges in maintaining grid reliability. Meeting these reliability challenges while effectively integrating the above resources requires a quantum leap in harnessing communication and information technologies. A common vision for cohesive integration of these technologies facilitates the convergence of standards and protocols that are so acutely needed and expedites the deployment of the technologies. Such a common vision can be reached through a systematic approach based on understanding of the reliability challenges and demand response in the modern power grid.

Demand response allows consumer load reduction in response to emergency and high-price conditions on the electricity grid. Such conditions are more prevalent during peak load or congested operation. Non-emergency demand response in the range of 5 to 15% of system peak load can provide substantial benefits in reducing the need for additional resources and lowering real-time electricity prices. Demand response does not substantially change the total energy consumption since a large fraction of the energy saved during the load curtailment period is consumed at a more opportune time to flatten the consumption profile of the customers.

Load rejection as an emergency resource to protect the grid from disruption is well understood and is implemented to operate either by system operator command or through underfrequency or undervoltage relays. In a smart grid, the load rejection schemes can be enhanced to act more intelligently and based on customer participation. Price-based demand response/load management as a system resource to balance demand and supply has not been widely adopted yet. Contract-based participation has been typically below 5% of peak load. In a smart grid, real-time price information enables wider voluntary participation by consumers. Demand response can be implemented through either automatic or manual response to price signals, or through a bidding process based on direct communications between the consumer and the market/system operator or through intermediaries such as aggregators or local utilities (Figure 6.2).

In addition to the capability to flatten the load profile, demand response can serve as an ancillary resource. As such, demand response schemes could improve reliability.

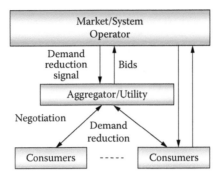

FIGURE 6.2
Communications for demand response (DR).

6.2.3 Demand Response-Based Architecture of the Smart Grid

The concept of the smart grid started with the notion of advanced metering infrastructure to improve demand-side management, energy efficiency and a self-healing electrical grid to improve supply reliability and response to natural disasters or malicious sabotage. However, several developments have led to the expansion of the initially perceived scope of the smart grid, and are helping shape the new face of the electricity industry. These include:

1. Emphasis on environmental protection, including renewable generation (wind, solar, etc.) and demand response (DR)
2. The drive for better asset utilization, including operating closer to the 'knee of the curve' (stability margin) while maintaining reliable system operation
3. Need for enhanced customer choice

Figure 6.3 schematically depicts these factors in relation to the new emerging smart grid paradigm, and illustrates the place of DR and more generally distributed energy resources (DERs) in the new arena.

Another emerging paradigm shift is the bi-directional interaction between wholesale markets/transmission operations and retail markets/distribution operations. The expected profusion of DR, renewable resources and distributed generation and storage at the distribution/retail level has direct implications on the operation of the transmission system and the wholesale energy markets. Enabling technologies, such as enhancements in the communication and information technologies, make it possible to turn these new resources into useful controllable products for wholesale market and transmission system operators. Figure 6.4 schematically shows the traditional utility environment in terms of the flow of power and information.

In contrast, in the emerging utility environment, both power and information flows are bi-directional, as shown in Figure 6.5. The emerging use of

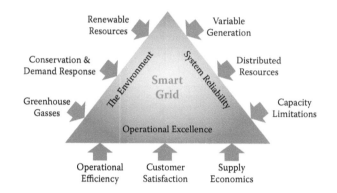

FIGURE 6.3
Industry drivers of the smart grid.

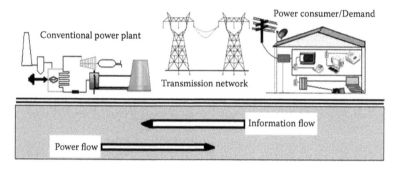

FIGURE 6.4
Power and information flow in traditional utility environment.

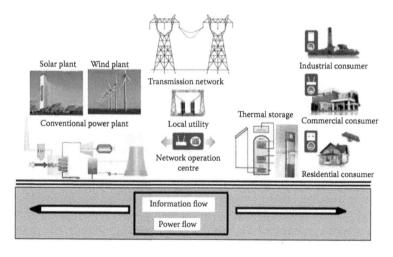

FIGURE 6.5
Power and information flow under the smart grid.

thermal storage for peak shifting, the anticipated growth and cost reduction of solar photovoltaic (PV) generation at residential and municipal levels, the anticipated shift from conventional fuel transportation to plug-in electric vehicles (PEVs), the advent of low-cost smart sensors and the availability of a two-way secure communications network across utility service territories are anticipated to significantly alter the nature of future power supply and power system operations, as well as consumer behaviour. DR is an important ingredient of the emerging smart grid paradigm and an important element in market design to keep the potential supply market power in check.

Hence, to utilize DR independent system operators (ISOs)/regional transmission organizations (RTOs) have to:

- Accept bids/offers from DR resources for ancillary services (A/S) comparable to any other A/S-capable resources (currently mostly generation resources)
- Allow DR resources to specify limits on frequency, duration and the amount of their service in bids/offers to provide A/S
- Impose charges to buyers for taking less electric energy in real time than scheduled, and eliminate such charges during system emergencies
- Permit aggregators to bid DR on behalf of retail customers directly into the market
- Study and report on reforms needed to eliminate barriers to DR in energy markets
- Assess, through pilot projects, the technical feasibility and value to the market of using A/S from small DR units

These provisions are being implemented at different ISOs/RTOs by modifying or expanding the existing market rules. Specific telemetry and control requirements may apply to DR resources for participation in some of these product markets.

Also, different ISOs/RTOs are adopting different rules for monitoring DR performance in relation to different products and for settlement with DR resource owners, operators and aggregators.

6.2.4 Effect of DR on the Smart Grid Scenario

Dispatchable DR management is the process of controlling end use customers (load). Both transmission and distribution system operators will have to deal with and benefit from DR actions. In ISO-operated markets, DR can provide information like installed capacity or unit commitment availability, day-ahead real-time balancing of energy, regulating spinning reserve/responsive reserve and supplemental reserve. The DR connectivity is shown in Figure 6.6.

FIGURE 6.6
Block diagram of DR connectivity.

Some more classes of market participants, including load serving entities (LSEs) and utility distribution companies (UDCs), can offer DR products to ISO/RTO to modify the day-ahead market. Hence, the ISO has to optimize the use of its utilities considering DRs, where the objective is not only to minimize the generation cost but also to maximize the utilization of DR for the most economical solution.

6.2.5 Cost Component of the Smart Grid

Included in the estimates of the investment needed to realize the smart grid, there are estimated expenditures needed to meet the continuously increasing consumer demand and to make possible large-scale renewable power production simultaneous with conventional power. As part of these expenditures, the components of the expanded power network will need to be compatible with the smart grid. Hence, the total costs to enable a fully functioning smart grid can be classified into three parts: (1) transmission and sub-station end costs, (2) distribution end costs and (3) consumer end costs.

In practice, more multi-faceted cost allocation methods might well be adopted. This calculation assumes that the smart grid costs are equalized over consumers across the country. However, the smart grid cost per consumer is likely to vary considerably, and therefore the total estimated smart grid cost may be more concentrated in some areas, which would raise the cost per consumer in those areas and reduce it elsewhere. These costs are modest when compared to the benefits the smart grid will yield. However, the challenge for all of those in the electricity sector will be communicating that the smart grid is indeed a good investment.

The higher costs of power in the smart grid are due to major elements (such as DR, renewable power sources, distributed generation, energy efficiency, energy savings corresponding to peak load management, etc.) of functionality added to the smart grid. However, smart grid component costs are declining rapidly. As these technologies mature and as production quantity increases, the marginal costs of smart grid technologies have the potential to decline rapidly.

6.2.5.1 Cost Components for the Smart Grid: Transmission Systems and Sub-Stations End

The main components of cost for the transmission and sub-station portion of the smart grid are as follows:

1. Transmission line sensors
2. Storage for bulk transmission wholesale services
3. Flexible AC transmission system (FACTS) devices and high-voltage direct current (HVDC) terminals
4. Short-circuit current limiters
5. Communications infrastructure to support transmission lines and sub-stations
6. Core sub-station infrastructure for IT and corresponding cyber-security
7. Electronics devices and phasor measurement technology for wide area monitoring
8. Outage management and distribution management, which may include faster real-time simulation, improved load modeling and forecasting tools, probable outage or vulnerability assessment, visualization and taking proper corrective or preventive action
9. Improved energy operation and sub-station automation

6.2.5.2 Cost Components for the Smart Grid: Distribution End

Smart grid investments in the distribution system require wider high bandwidth communications to all sub-stations, intelligent electronic devices (IEDs) that provide flexible control and protection systems, complete distribution system monitoring that is integrated with larger asset management systems, including dynamic sharing of computational resources of all IEDs, and distributed command and control to mitigate power quality events and improve reliability and system performance. The key cost components for the distribution portion of the smart grid are as follows:

1. Communications between all digital devices on the distribution system
2. Distribution automation

3. Distribution feeder circuit automation that may include intelligent auto-reclosers and relays, power electronics, including distribution short-circuit current limiters, voltage and VAR control on feeders
4. Intelligent universal transformers
5. Advanced metering infrastructure (AMI)
6. Local controllers in buildings, on microgrids, or on distribution systems for local area networks

6.2.5.3 Cost Components of the Smart Grid: Consumer End

The key components for the customer portion of the smart grid costs are listed below:

1. Integrated inverter for PV adoption
2. Consumer residential energy management system (EMS) portal and panel
3. In-home displays
4. Grid-ready appliances and devices
5. Vehicle-to-grid two-way power converters
6. Residential storage for backup
7. Industrial and commercial storage for power quality
8. Commercial building automation

6.2.6 Smart Grid: Cost-Benefit Analysis

In the context of cost-effectiveness of smart grid reliability investments, it is essential for utilities and regulators to have a universal framework for cost-benefit analysis that properly accounts for the communal benefits that arise from utility investments in reliability. Basically, a smart grid bears a resemblance to the computer. Sensors in various locations on the grid gather information on system operating conditions and transmit that information (more or less instantaneously) to utility computers to make decisions and any required changes to grid equipment settings without human involvement, continuously. In many cases, these changes can proactively address issues before they create problems for both the power producer and consumers. Information can also be stored for future analysis of system performances. To develop these communication circuits, the power investor requires a large investment, but it can be easily optimized with the benefits that can be provided by the smart grid.

- The smart grid reduces carbon emissions and enhances energy efficiency through reduction in losses and by integrating renewable generation. By flattening the consumption profile, this system can run much more efficiently and can respond more intelligently, with its potential.

- The smart metres can provide general information on the nature and extent of service outages and fault location, whereas in a deregulated power network, the ISO will analyze the locations of outages or try to narrow down the location of a fault to a particular distribution or transmission line. The repair team then takes action to clear the fault. But, for underground cables, the fault detection mechanism is somehow different, and here the repair team must physically examine multiple pieces of equipment to identify locations by a process of elimination. All of these efforts take a lot of time.

- In a smart grid, several types of devices on a distribution line can serve to isolate a section of distribution line on which a fault has occurred. These devices, generally called sectionalizing devices, operate automatically by sensing a reduction in electric parameter (such as over/undervoltage, current or frequency, etc.).

- Smart meter and other intelligent devices present in a smart grid can help to manage the challenges offered by the integration process of renewable energy with conventional power, thereby increasing the amount of renewable generation that can be reliably and efficiently integrated.

- The bi-directional nature of energy and information flow in a smart grid allows consumers to connect their own generation, such as solar panels, to the grid. This adds flexibility and reduces greenhouse gases while putting less stress on the grid system and thus reducing losses. But, the economic benefit of accommodating increasing levels of renewable generation with conventional power may raise indecision. There are increased costs associated with renewable generation in the short term, including the investments required to accommodate it with the higher capital investment required to build it. On the other hand, there are economic advantages to renewable generation over the long term, including the avoidance of fuel costs and the potential economic consequences associated with rapid climate disruption. Thus, the smart grid provides long-term socio-economic and cost-effective quality power.

From the overall discussion about the smart grid, one may conclude that in actuality, the increased costs are a reflection of a newer, more advanced vision for the smart grid. The concept of the base requirements for the smart grid is significantly more expansive, but its benefits allow power investors with large investments, which in the long run provides more cost-effective benefits for both the power producer and consumers.

Hence, in the following section, effort has been made to develop different utility-optimized methodologies under the smart grid environment.

6.3 Swarm Intelligence-Based Utility and Cost Optimization

The demand response-based resource management and load scheduling has proven to be one of the key topics for computing in the smart grid environment. In fact, viewed from the transmission and wholesale operations in the case of the energy and ancillary services market facilitated by the ISO, the DR can improve the system operating conditions.

Henceforth, the objective of the work discussed in this section is to optimize the generation and load schedule with maximum utilization of DR information [312]. As in the smart grid, it is expected that the constraints of the power network will increase to its optimum value; the objective function will be highly non-linear, which has compelled us to choose a stochastic optimization technique like swarm intelligence.

6.3.1 Cost Objective and Operating Constraints of the Work

From a social fairness point of view, it is desirable to utilize the available capacity provided by the energy provider in such a way that the market price variation is minimal with the increase in demand or with the variation of DR. However, as the DR is provided by end users, they will have the option to choose their own consumption level to maximize their own welfare. To align individual welfare with social welfare the objective function has been developed as follows:

$$\text{Minimize} \quad \frac{\left(\sum_{g=1}^{Ng} C_{TD} - \sum_{g=1}^{Ng} C_T \right)}{\sum_{i=1}^{LB} d_i} \tag{6.1}$$

Equality or power balance constraints:

$$\sum_{i=1}^{NG}\sum_{g \in G}\left(P_{gi}^0 + \Delta P_{gi}^j \right) - \left(P_i^0 + \sum_{i=1}^{NB} \Delta P_{di} - LCP^j \right) = \sum_{k=1}^{N} \left| V_i^j \right| \left| V_k^j \right| \left| Y_{ik}^j \right| \cos\left(\theta_i^j - \theta_k^j - \delta_{ik}^j \right) \tag{6.2}$$

$$\sum_{i=1}^{NG}\sum_{g\in G}\left(Q_{gi}^{0}+\Delta Q_{gi}^{j}+\right)-\left(Q_{li}^{0}+\sum_{i=1}^{NB}\Delta Q_{di}-LCQ^{j}\right)=\sum_{k=1}^{N}\left|V_{i}^{j}\right|\left|V_{k}^{j}\right|\left|Y_{ik}^{j}\right|\sin\left(\theta_{i}^{j}-\theta_{k}^{j}-\delta_{ik}^{j}\right)$$

(6.3)

Inequality or generator output constraints:

$$P_{Gi}^{\min}\le P_{Gi}^{0}\le P_{Gi}^{\max}$$

(6.4)

$$Q_{Gi}^{\min}\le Q_{Gi}^{0}\le Q_{Gi}^{\max}$$

(6.5)

$$\Delta P_{Gi}^{\min}\le\Delta P_{Gi}^{j}\le\Delta P_{Gi}^{\max}$$

(6.6)

$$\Delta Q_{Gi}^{\min}\le\Delta Q_{Gi}^{j}\le\Delta Q_{Gi}^{\max}$$

(6.7)

Voltage constraint: $\left|V_{i}^{\min}\right|\le\left|V_{i}^{j}\right|\le\left|V_{i}^{\max}\right|$

(6.8)

Transmission constraint: $P_{ij\,\min}\le P_{ij}\le P_{ij\,\max}$

(6.9)

Load curtailment limits: $0\le LCP^{j}\le P_{i}^{\max}$

(6.10)

where C_{TD}, C_{T} are the total costs of generation with the change of DR and constant demand, respectively; b_{i}, d_{i} are the bidding price and demand offered by the ith load; 0, j are the normal and contingency state indices (superscript); P_{ij}^{0} & P_{ij}^{j} are the line flows in MW between bus i and bus j before and after tripping; P_{i}^{0} & P_{i}^{j} are the total powers injected to bus i before and after tripping, $P_{ij\,\max}$ is the maximum line flow to be allowed between the ith and the jth bus; P_{gi}^{0}, Q_{gi}^{0} are the real and reactive powers dispatched; LCP^{j}, LCQ^{j} are the real and reactive power load curtailments in MW and MVAR; $\left|V_{i}^{j}\right|\angle\theta_{i}^{j}$ is the bus voltage in p.u.; and ΔP_{di}, ΔQ_{di} are the changes in active and reactive power demand of the ith load as informed by DR.

In the smart grid environment, the loads are to be classified in such a way that they receive minimum curtailment with their proposed price or bidding price of electricity. For this classification considering both present market conditions and willingness to pay, an attribute has to be added with each load, which can act as a DR monitor in variable demand conditions or any contingency. With the assistance of this new attribute or index the ISO will be able to produce an optimum generation and load schedule, which can be referred to consumers for their consideration to compromise with curtailment or to increase their offerings.

6.3.2 Theory of Cost-Regulated Curtailment Index (CI)

The market demand originates from what the participants are willing to pay. It starts with the bidding price of the consumers (DISCOs), and their willingness to pay is determined by calculating the difference between the bidding price and the market clearing price (MCP). For the curtailment of load this willingness to pay is considered, and the consumer with the lowest value of the same will suffer a higher degree of curtailment. For the standardization of this rate of curtailment process, a monitoring index is required that can trace all the loads according to their willingness to pay.

For the development of such an index the rate of generation cost (RGC) for a particular load is required, which can be written as

$$RGC = \frac{\sum_{i=1}^{NG} \alpha_i P_i^2 + \beta_i P_i + \gamma_i}{\sum_{i=1}^{LB} d_i} \qquad (6.11)$$

For the ith load,

$$\text{Generation cost to cater the demand} = RGC * d_i \qquad (6.12)$$

The sustainability to supply to the ith load can be ensured if and only if the consumer pays this cost at a breakeven point. Thus, for a bidding price b_i the condition for retaining the load is given by

$$b_i \geq RGC * d_i \qquad (6.13)$$

This inequality has been plotted as shown in Figure 6.7. The area of solution as stated should lie to the left-hand side shown by shaded lines. A flat bidding curve has been assumed that may alter according to the willingness of the consumer to pay. The profit margin increases as the shaded portion to the left of the intersecting point increases. But, for social welfare in terms of optimum market price, it can be achieved by confining the solution area near the marginal condition.

Thus, the marginal condition for optimum market price variation is

$$b_i = RGC * d_i \qquad (6.14)$$

Hence, the ratio of the bidding price to the generation cost of a particular load should be unity owing to the marginal condition. Thus, the curtailment index (CI) has been developed as below:

$$CI = \frac{RGC * d_i}{b_i} \qquad (6.15)$$

It is worth mentioning that for topological and socio-economic reasons, both the generation cost-demand curve and bidding price-demand curve can be non-linear. As in this work, the marginal condition is the only point of excursion; the curves have been assumed to be linear.

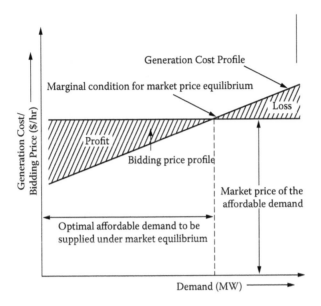

FIGURE 6.7
Graphical representation of marginal price at equilibrium.

6.3.3 Cost Realization Methodology Implementation with Swarm Intelligence

With the objective function formulated in (6.1) the flowchart depicted in Figure 6.8 presents the methodology adopted in a swarm optimization environment for utility maximization in a smart grid with DR. As the methodology considers DR and social welfare simultaneously, the optimization algorithm requires a convex formulation of the objective function. The classical lambda iteration process of optimization may fail to provide a global solution, and hence PSO has been chosen for its high degree of reliability, accuracy and speed, which are essential in the smart grid environment.

In the following methodology, whenever the value of developed CI exceeds unity for a particular load, the consumer is requested to reduce his demand or increase his bidding price to survive in the market. To accelerate the convergence of the optimization, an acceleration factor (AF) has also been developed (6.16) and utilized that assists in curtailment.

$$AF_i = \frac{\Delta\, Load\, curtailment_i}{\Delta\, Demand_i} \tag{6.16}$$

A higher value of AF will suggest a faster convergence of the OPF, but low accuracy of curtailment. Hence, a suitable value of AF should be chosen.

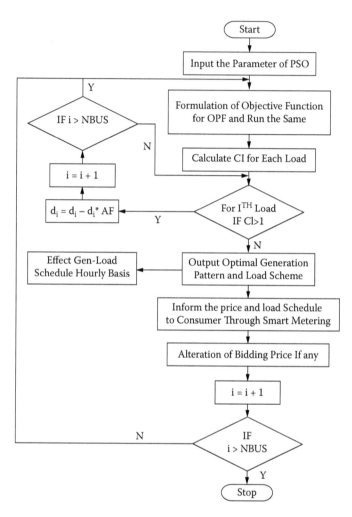

FIGURE 6.8
Flowchart of the developed methodology.

On an hourly basis the results of the developed methodology can be put into effect, but the mainframe algorithm should run at a high frequency so that the DR and consumer participation can increase with smart metering. As shown in Figure 6.8, for real-time participation of consumers the load schedule should be updated at a faster rate with regard to the generation schedule. The price information with advanced metering technology can reach the end user with optimum ease, and proper utilization of this methodology will be able to revive the consciousness of the consumers against misuse of electricity. Hence, the technological advancement in this way not only contributes to the most economic operation, but also will enforce regulation against unscheduled uses of electricity.

6.3.4 Implementation of the Cost-Effective Methodology with DR Connectivity

The effectiveness of the developed algorithm has been presented in a test system (IEEE 30 bus) with smart metering facility. The OPF result with constant demand is depicted in Table 6.1, showing the total generation cost and the p.u. generation cost on an hourly basis. From this table the minimum cost of dispatch for each consumer can be enumerated. In the base case, it has been assumed that the market is operating at equilibrium; that is, every load is contributing at the rate shown in Table 6.1. The causal bidding prices, however, are depicted in Table 6.2. As the market is operating at equilibrium, the bidding price of each customer has been assumed as the product of p.u. generation

TABLE 6.1

Generator Schedule for Unity CI

P_1 (MW)	P_2 (MW)	P_3 (MW)	P_4 (MW)	P_5 (MW)	P_6 (MW)	Total Gen Cost $/h	p.u. Gen Cost $/MWh
176.8	48.82	21.45	21.60	12.08	12.00	801.8	2.83

TABLE 6.2

Projected Bidding of LDCs in Market Price Equilibrium

Load Connected at Bus	Demand in MW	Bidding Price by the Consumer ($/h)
2	21.70	61.39
3	2.40	6.79
4	7.60	21.50
5	94.20	267.37
7	22.80	64.50
8	30.00	84.88
10	5.80	16.41
12	11.20	31.68
14	6.20	17.54
15	8.20	23.20
16	3.50	9.90
17	9.00	25.46
18	3.40	9.61
19	9.50	26.87
20	2.20	6.22
21	17.50	49.51
23	3.20	9.05
24	8.70	24.61
26	3.50	9.90
29	2.40	6.79
30	10.60	29.99

cost and its individual demand. Thus, the developed CI of each load remains at its ideal value of unity. The bidding price, however, is incessant and subjected to market variation; i.e. it can change by the next hour.

When the demand in the market increases, the generation cost, and hence market price, also increases due to redispatching. This market price spike can be reduced by proper selection of load for curtailment. In the algorithm the curtailment index assists for the purpose. If CI exceeds unity for a particular load, it would suggest that the price of electricity for that load is greater than it can afford. Hence, the algorithm traces all the loads and calculates the value of this index for each of them and continues the same until and unless CI for all the loads gets to or below unity in real time, and in the process it reduces the price volatility of the market. To testify to the applicability of the algorithm in the smart grid environment, the demand of bus 5 has been increased by 10, 15 and 20%, respectively. As the swell in load has been concentrated on bus 5, the optimized value of CI could not reach unity (Table 6.3). Hence, to restore market equilibrium, bus 5 has to pay the penalty for the above rise in demand. It is worth mentioning that in case of the conventional controlled load curtailment technique (frequency or voltage stability based), the consumer with a demand hike has only to suffer curtailment or comparatively more penalties for real-time restoration of the frequency or voltage profile of the system.

Apart from restoring the market equilibrium in a smart grid environment, the algorithm can prepare a modified load schedule according to bidding of different consumers. In the smart grid, different types of consumers may have different preferences. Some users may desire a low cost of energy, and some may opt for sustainability of the same offering high price. The algorithm can effectively segregate the consumers according to their willingness to pay for sustainability. The curtailment index (CI) goes below unity with a higher bidding price. Thus, the load schedule prepared by this algorithm is the most feasible schedule, and the consumers have to reduce their consumption, as they desire a minimum price more than sustainability. The ISO, on application of this algorithm, can send this price information via smart metering to the consumers to either reduce their consumption according to the load schedule obtained or increase their bidding price for maintaining the former contract. The load schedule obtained with 10, 15 and 20% uniform increase in demand by the consumer connected in bus 5 is given in Table 6.4. As observable from the table, the load curtailment is not concentrated only in the region, which is responsible for the market in equilibrium; rather, the curtailment is a distributed pattern that identifies the loads with their bidding and subjects them to a periodic load reduction.

So far the study has focused on the effectiveness of the developed algorithm with a concentrated rise of demand at a particular bus. The assessment of the applicability of the developed method remains incomplete without testing its capability of restoring market equilibrium in variable distributed demand conditions (refer to column 3 of Table 6.5).

TABLE 6.3

Improvement of the Value of CI with the Developed Algorithm

Load Bus	CI for 10% Increased Demand with		CI for 15% Increased Demand with		CI for 20% Increased Demand with	
	Conventional Cost Optimization	Developed DR Algorithm	Conventional Cost Optimization	Developed DR Algorithm	Conventional Cost Optimization	Developed DR Algorithm
2	1.03	0.95	1.047	0.92	1.06	0.89
3	1.03	0.95	1.047	0.92	1.06	0.89
4	1.03	0.95	1.047	0.92	1.06	0.89
5	1.34	1.10	1.515	1.16	1.69	1.21
7	1.03	0.94	1.047	0.92	1.06	0.89
8	1.03	0.94	1.047	0.92	1.06	0.89
10	1.03	0.94	1.047	0.92	1.06	0.89
12	1.03	0.94	1.047	0.92	1.06	0.89
14	1.03	0.94	1.047	0.92	1.06	0.89
15	1.03	0.94	1.047	0.92	1.06	0.89
16	1.03	0.94	1.047	0.92	1.06	0.89
17	1.03	0.94	1.047	0.92	1.06	0.89
18	1.03	0.89	1.047	0.86	1.06	0.89
19	1.03	0.94	1.047	0.92	1.06	0.89
20	1.03	0.94	1.047	0.92	1.06	0.89
21	1.03	0.94	1.047	0.92	1.06	0.89
23	1.03	0.94	1.047	0.92	1.06	0.89
24	1.03	0.94	1.047	0.92	1.06	0.89
26	1.03	0.94	1.047	0.92	1.06	0.89
29	1.03	0.94	1.047	0.92	1.06	0.89
30	1.03	0.94	1.047	0.92	1.06	0.89

TABLE 6.4

Load Schedule Offered by Developed Algorithm in Variable Condition of Demand

Load Connected at Bus	Proposed Optimized Load Schedule after		
	10% Increase in Demand	15% Increase in Demand	20% Increase in Demand
2	20.53	19.92	19.30
3	2.27	2.20	2.13
4	7.19	6.97	6.76
5	103.59	109.04	114.27
7	21.57	20.92	20.28
10	5.48	5.32	5.16
12	10.6	10.27	9.96
14	5.86	5.69	5.51
15	7.76	7.52	7.29
16	3.31	3.21	3.11
17	8.51	8.25	8.00
18	3.02	2.93	2.84
19	8.99	8.71	8.45
20	2.08	2.01	1.95
21	16.56	16.06	15.57
23	3.03	2.93	2.84
24	8.23	7.98	7.74
26	3.31	3.21	3.11
29	2.27	2.20	2.13
30	10.03	9.72	9.43

The CI has to monitor the changes in demand in all cases, and the described algorithm with smart metering has to prepare the most reasonable genera-tion and load schedule. The added advantage of this algorithm is, along with the generation schedule, it prepares a formidable load schedule, showing each load status by fast metering in the smart grid and keeping consum-ers alert about their eligibility of staying in the market with their demand. Table 6.5 depicts the variable demand and the corresponding DR with the new algorithm.

The operational loss reduces considerably with the application of this algorithm. Reduction of line losses instigates the rise of system performance in terms of efficiency. The conventional curtailment-based algorithm can-not flatten the line loss profile as can be done in the case of the developed algorithm (Figure 6.8). The flat loss profile also suggests that this algorithm is capable of limiting market price variations in the smart grid. Figure 6.9 shows the effective improvement of the line loss profile with respect to the conventional algorithm.

The efficacy of this algorithm is quite evident in terms of price volatility minimization. In the conventional method, to sustain the price of electricity,

TABLE 6.5

Restoration of Market Equilibrium in Distributed Increase in Demand

Bus No.	Load (MW)	Change of Demand Profile in DR Centres Submitted by Consumers	CI for 10% Increased Demand with Conventional Optimization	Proposed Optimized Load Schedule after Curtailment	CI for 10% Increased Demand with the New DR Algorithm
2	21.7	15% increase	1.028277	21.65569	0.99821
3	2.4	2% increase	1.028277	2.31837	0.96618
4	7.6	6% decrease	1.028278	7.14400	0.94026
5	94.5	10% increase	1.028275	93.87136	0.99350
7	22.8	2% decrease	1.028278	21.16080	0.92836
8	30	25% increase	1.028277	32.18260	1.07888
10	5.8	2% decrease	1.02828	5.38301	0.92883
12	11.2	6% decrease	1.028278	9.97050	0.89115
14	6.2	Constant	1.028277	5.87169	0.94788
15	8.2	Constant	1.028278	7.76578	0.94780
16	3.5	10% increase	1.028276	3.47672	0.99438
17	9	1% increase	1.028278	8.60865	0.95741
18	3.4	Constant	1.028277	3.21996	0.94874
19	9.5	Constant	1.028278	8.99694	0.94808
20	2.2	2% increase	1.028276	2.12517	0.96744
21	17.5	50% increase	1.028276	22.52782	1.29556
23	3.2	10% increase	1.028276	3.17871	0.99498
24	8.7	Constant	1.028278	8.23930	0.94840
26	3.5	2% decrease	1.028276	3.24837	0.92949
29	2.4	3% increase	1.028277	2.28785	0.95449
30	10.6	Constant	1.028278	10.03869	0.94823

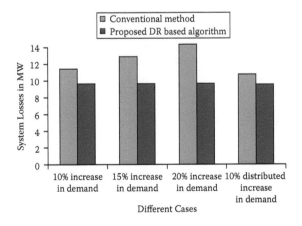

FIGURE 6.9
Improvement of line losses.

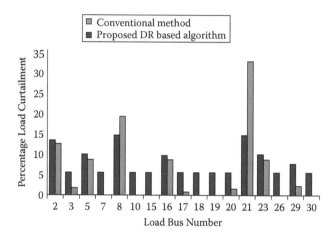

FIGURE 6.10
Comparison of load curtailment between developed and conventional methods.

the curtailment is effected in the buses, where the load increase is more prudent, disregarding the willingness to pay of the loads. This will affect the reliability of power supply, and the consumers will be discouraged from paying more than the MCP for sustainability of power supply. In the developed algorithm, with more distributed curtailment, every consumer will be encouraged to escalate its bidding price to retain its load. The percentage curtailment hence is more concentrated on specific loads with the conventional algorithm; on the contrary, this curtailment is distributed in case of the developed algorithm, as shown in Figure 6.10.

For on-line application of the developed algorithm, the convergence characteristics have been examined, and it has been found, as shown in Figure 6.10, that it is taking almost the same number of iterations to restore market equilibrium for different rates of increase in demand. In the smart grid, after each of these sets of iterations consumers can be fed with the information regarding curtailment adoption for market price volatility minimization.

6.4 Summary

There is a pressing requirement to accelerate the development of low-carbon energy, cost-effective technologies in order to address the global challenges of energy security, power quality, reliability and economic growth. The smart grid is particularly the key solution as it coordinates the needs and capabilities of all generators, grid operators and power consumers to operate all parts of the system as efficiently as possible by minimizing costs and environmental impacts while maximizing system reliability, resilience and stability using the demand response concept.

Demand response (DR) can be defined as changes in electric usage by power consumers from their normal consumption patterns in response to changes in the price of electricity over time during a full day. In the conventional cost optimization technique the DR of the consumers is not incorporated, and hence the ISO has to rely upon the forecasted load, which at times may cause inefficient operation of the system. A smart grid is, however, endowed with the facility of DR, which can be effectively used to increase the efficiency of the system. In this context, the described methodology presents a price volatility optimization methodology capable of assessing DR and the willingness to pay factor in real time by tracing each load for its competency to retain its place in the market without curtailment. This technique classifies the loads in terms of willingness to pay and according to DR selection. It also reschedules the generation of GENCOs as well as the load profile to restore power market price equilibrium. It has been depicted that the implementation of the methodology in the smart grid can effectively improve the reliability of the system and also maximize social welfare.

Annotating Outline

- The world's electricity systems face a number of challenges, including ageing infrastructure, continued growth in demand, the integration of variable renewable energy sources, the need to improve the security, reliability and quality of supply and the need to lower carbon emissions. The smart grid may be the answer to the uncertainties and currently faced challenges offered by existing power networks.

- Due to the emergence of advanced metering infrastructure and communication protocols, the next era of power systems will belong to the smart grid.

- The smart grid can be characterized by self-healing capability against external disturbance. It also has potential to enable active participation of consumers, to accommodate all generation and energy storage technologies and to optimize asset utilization and efficiency.

- Improvement of performance with the smart grid is subject not only to the preparation of a generation schedule but also to manifest consumer participation.

- The DR technology incorporated in the described algorithm has increased the involvement of consumer bidding in retaining the market price equilibrium.

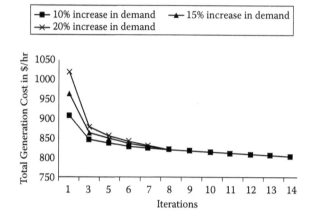

FIGURE 6.11
Convergence curves of OPF.

- The algorithm, with the assistance of the developed CI, effectively monitors all the loads with their bidding and in case of any inadvertent conditions, it produces an equitable contribution to maintain the equilibrium in terms of price of electricity (Tables 6.3 to 6.5) and line loss (Figure 6.9).

- To maintain better control on the optimization process for real-time operation an AF has also been developed that, apart from minimizing the curtailment, effectively contributes to synchronizing the generation schedule, load schedule and bidding.

- The methodology described has been demonstrated to be fast (Figure 6.11), accurate and cost-effective incorporating DR in the smart grid arena for better optimization and utilization of the existing grid.

- Thus, the smart grid will come from the application of cost-effective, intelligent energy technology to optimize the use of generation resources and the delivery of power.

7

Epilogue

The main results from the different chapters together with some general conclusions have been accumulated in this chapter. Finally, some potential extensions of this effort are proposed.

7.1 Summary and Conclusions

This book has dealt with cost effective multi-objective methodologies dedicated for assessment, control and improvement of power network performance in normal, stressed and contingent states. The recent evolution of the structure of the power network has significantly increased the interest in this matter, and it has become an issue of great concern for utility companies, system operators and consumers. Traditional methodologies for improving system performance in terms of voltage stability, power flow allocation and efficiency in vertically integrated environments used to work moderately but were universally accepted in the absence of any other alternatives. Due to the enormous alteration of power market operation with the introduction of deregulation and smart metering, the conventional methodologies can no longer be reliable and should immediately be modified to adopt the paradigm shift of the system. The negative impacts of the present power network with conventional techniques have reached an alarming state, and in the new arena of embedded generation and unbundling of the electric power industry, the shift cannot be prolonged.

On the other hand, in most of the power markets around the globe frequency-based stability restoration and system operation are followed, which is of course simple and less costly but runs the system in sub-optimal mode. The environmental issues are also restricting the conventional methods to reach the optimal solution. Hence, investments in terms of procurement of modern technologies like flexible AC transmission system (FACTS) and high-voltage direct current (HVDC), and appropriate employment of optimization methodologies have become quite imperative for the system and challenging for power engineers.

A comprehensive literature survey of the present power market scenario and performance improvement methodologies has been included in Chapter 2. Some serious issues regarding the present and proposed future power network have been discussed in detail starting from voltage stability assessment methodologies up to the multi-objective optimization techniques, highlighting the utilization of FACTS and HVDC technologies. The survey continued in different directions, originating with an objective of improving network performance in terms of static operation of the power network subject to various steady-state disturbances. After enlightening the difficulties of assessing stability problems with classical techniques, the possibility of artificial neural network (ANN)-based methodologies was studied. To accomplish a wide stability margin, the implications of FACTS and HVDC devices were also discussed. Efficient methodologies for optimum operation of modern deregulated power markets have been the point of focus in the last decade. Though it has not yet been completely enforced, the literature survey conducted illuminates the techniques for the improvement of its performance by minimizing cost volatility and line congestion. The availability of efficient smart metres and challenges of the smart grid have also motivated the authors to pay attention to the recent development in this area of the power system.

Voltage stability is an essential requirement in industry as well as in the transmission sector, and recently, it has drawn even more attention in a competitive energy market under a deregulated environment. Chapter 3 presents novel approaches to recognize instability in classical as well as neo-classical methods. In the preliminary stage, the classical voltage stability indices dedicated to identify static voltage instability or point of collapse have been simulated for a multi-bus power network, and the results obtained have been compared in tabular form to indicate the applicability of these static indices in off-line studies. A vivid layout of ANN and its applicability in faster recognition of voltage stability has also been demonstrated. Some promising results of the ANN-based stability model have been presented in this chapter to establish the supremacy of the model over the classical approach. As an added advantage, this developed ANN-based model does not initiate any price volatility in the power network, as it does not entail any additional investment by the power investor.

Improvement of system performance in terms of voltage stability and power handling capability with HVDC and FACTS devices with a focus on their installation cost has been depicted in Chapter 4. But, this method calls for additional investment to serve the power consumer properly. To establish the benefits of HVDC and FACTS devices in the long run, this chapter essentially regrooves the requirements of modern power networks in stressed and contingent states and places the compensating device at weak buses to enrich the performance of the system. It can be revealed from the study of this chapter that the effective utilization of these modern technologies can reconcile modern-day power network issues.

The deregulated power markets have been designed to harmonize the requirement of modern-day power system operators and consumers.

It is likely to be enforced in all the power markets around the globe in the middle of this decade. In deregulated power markets, there is an opportunity to take care of consumer welfare in terms of power quality, reliability and security. The excess monetary investment for maintaining power quality and security may be able to justify the price volatility in the power market. But, the process of deregulation remains incomplete unless the power market operators can standardize a common objective. Though the objectives like cost minimization, loss minimization and congestion management are common, their infusion in a single objective is a mammoth task considering the inextricable complex constraints. Again, due to the versatile nature of resources and diverse requirements of consumers, the formulation of these functions gets more complex unless multi-objective optimization algorithms are not being thought of, to maximize the proper distribution of available resources. In this pursuit, Chapter 5 has demonstrated some methodologies in stochastic optimization environments for real-time monitoring and control of social welfare with respect to cost volatility minimization with optimum efficiency. For this purpose, the authors have relied on neo-classical optimization techniques rather than classical optimization methods, as the modern power system constraints have exceeded the handling capability of conventional methods.

It has been established that the work presented in the genetic algorithm (GA)-based transmission loss optimization technique can effectively redistribute the generation pattern for maximizing the efficiency of the system. It has also been depicted that the effective utilization of multi-objective GA-based optimization can even limit line congestion. In the same pursuit of the search for a better solution, the authors have even implemented particle swarm optimization (PSO) for congestion management. The most highlighted part of this work is that the effective utilization of these algorithms can turn up with a single solution to loss minimization and congestion management, and thus both the objectives have been achieved in a single optimal solution.

With the goal of limiting price volatility, differential evolution (DE)- and PSO-based optimization techniques have been demonstrated. In the continuation of progress of the work, the authors have felt the need for proper assessment of the importance of a particular optimization technique. To this end, the authors have effectively formulated a few indices to determine the adoptability of a particular multi-objective solution in a power market and tested it in a swarm intelligence-based contingency surveillance technique. The demonstration and recommendations made in this work for the choice of the proper multi-objective optimization algorithm will have a positive impact on the performance of present-day power networks.

The chapter is thus an attempt to respond to the need for real-time cost-effective multi-objective optimization methodologies for modern power systems. Without defying the operating constraints, the multi-objective algorithms designed in this chapter have proven to be useful to generate solutions for modern-day power network problems.

There are two recent inclinations regarding what may become a paradigm shift in the way that power systems are designed and operated: (1) inclusion and rapid growth in renewable energy systems as sources of power supply, and (2) the modernization of industrial and domestic equipment and the increase in the use of computers, server farms, data banks, electronics and other non-linear loads. These trends are increasing the need for an intelligent grid concept that relies on communication between independent system operators/regional transmission organizations and consumers so that renewable energy can be used more efficiently, the electric power system can run more reliably and greenhouse gases can be reduced. For this purpose, recent development in the field of instrumentation and measurement has evolved with smart metering technology, and the power markets are supposed to adopt this technology in the near future, to be referred to as the smart grid. Chapter 6 comprises a basic portrayal view of the smart grid with its different components. It is awfully apparent that the inclusion of different intelligent devices and other facilities with the smart grid calls for more investment. Consequently, a brief cost-benefit analysis is also put together with this chapter. Thus, to maintain power producer and consumer welfare, few stochastic optimization techniques in the smart grid for the optimization of utility management have been discussed. As the smart grid continues to develop, and demand-side management, which is available today, builds on its growing reputation as a cost-effective way for industrial users to manage energy usage and costs, buy-in from both residential and industrial consumers will become simpler. There is minimal cost to get these facilities. The communicating network requires the simple installation of metres to monitor equipment more closely. By monitoring the equipment, smarter decisions regarding electricity usage can be made. The next step, connecting a network of loads to a regional electricity system operator, allows companies to respond to the needs of the grid and start offering grid balance. This creates a new revenue stream for companies with connected loads.

The work described in this chapter effectively segregates the loads according to their willingness to pay and produces generation and load schedules to increase the reliability of the system. The added advantage of the developed algorithm, along with the generation schedule, prepares a permissible load schedule, showing each load status by smart metering, and thus keeping the consumer alert about its eligibility of staying in market with its demand condition.

Given the above results, this book thus presents a new generation of highly flexible, accurate and highly capable multi-objective methodologies applicable in real time for addressing the complicated modern-day power network issues.

7.2 Future Scope

During the course of this study, different issues have been detected, among which a few are listed here as possible topics for future work in this area:

- As the maintenance of voltage stability is the most important issue, the artificial intelligence (AI)-based stability assessment models can be extended to be very effective in real-time monitoring and control of the bus voltage profile. In this pursuit, the modeling of FACTS and HVDC devices developed in this work can assist the system operator (SO) to incorporate these modern technologies for the improvement of system performance in the long run from an economy point of view.

- The GA-based loss optimization technique can be extended to incorporate emission coefficients and pollution constraints to emphasize the production of green energy. Again, the GA-based congestion management methodology can be converted to the congestion allocation methodology to effectively utilize the transmission lines of different ratings.

- DE- and PSO-based multi-objective solutions for cost volatility minimization can be used to infuse cost of reliability and cost of power quality in its objective function, and in due course, the developed optimization techniques can be utilized to incorporate renewable energy sources to encourage local generation, and thus system performance can be improved.

- The demand response (DR)-based algorithm can effectively include self-healing capability during contingency by presenting an hourly based load schedule. The work can be extended to include reliability and power quality issues to prepare a formidable load schedule to offer reliability and power quality as a product to be purchased in inadvertent states of the system.

- Smart grid technologies show strong potential to optimize asset utilization by shifting peak load to off-peak times, thereby decoupling electricity growth from peak load growth. A future work can be extended to present an ongoing effort to develop a peak load modeling methodology and estimate the potential of smart grid technologies.

- The essential requirements of a modern power network are generating the most economical schedule for social welfare with standard operational status (such as voltage profile, reliability and optimum load distribution) and maintaining non-volatility in electricity pricing after the procurement of an optimal solution. Work can be extended to amalgamate these two essential necessities of modern power markets, and hence to develop a new methodology to obtain the most economically feasible and reliable operating condition with the assistance of a new curtailment approach, and also make an effort to reduce oscillations of pricing signals in case of contingencies and intermittency of generating sources.

- A number of technical and measurement challenges to utility-scale integration of renewable energy sources are present, among which one of the major issues is the lack of adequate modeling tools for operations and planning for the use of bulk renewable resources. In this regard, a modern methodology can be developed that will fully optimize the use of renewable sources at utility scale.

References

1. Z. Alaywan, J. Allen, California Electric Restructuring: A Broad Description of the Development of the California ISO, *IEEE Transactions on Power Systems*, 13(4), 1445–1452, 1998.
2. F. Wu, P. Varaiya, P. Spiller, S. Oren, Folk Theorems on Transmission Access – Proofs and Counterexamples, *Journal of Regulatory Economics*, 10(1), 5–23, 1996.
3. New York ISO, *NYISO Manual for Transmission Services*, September 1999.
4. S. C. Srivastava, P. Kumar, Optimal Power Dispatch in Deregulated Market Considering Congestion Management, in *International Conference on Electric Utility Deregulation and Restructuring and Power Technologies*, April 2010, pp. 53–59.
5. D. J. Sobajic, Y.-H. Pao, Artificial Neural Net Based Dynamic Security Assessment for Electric Power System, *IEEE Transactions on Power Systems*, 4(1), 220–228, 1989.
6. L. Wang, M. Kiein, S. Yirga, P. Kundur, Dynamic Reduction of Large Power System for Stability Studies, *IEEE Transactions on Power Systems*, 12(2), 889–895, 1997.
7. E. Vahedi, Y. Mansoor, C. Fuches, M. Lotero, Dynamic Security Constraint of the Optimal Power flow/Var Planning, *IEEE Transactions on Power Systems*, 16(1), 38–43, 2001.
8. M. Stubbe, A. Bihain, J. Deuse, Simulation of Voltage Collapse, *Electrical Power and Energy Systems*, 15(4), 1993.
9. J. Deuse, M. Stubbe, Dynamic Simulation of Voltage Collapses, *IEEE Transactions on Power Systems*, 8(3), 894–904, 1993.
10. G. K. Morison, B. Gao, P. Kundur, Voltage Stability Analysis Using Static and Dynamic Approaches, *IEEE Transactions on Power Systems*, 8(3), 1159–1171, 1993.
11. J. Deuse, M. Stubbe, Dynamic Simulation of Voltage Collapses, *IEEE Transactions on Power Systems*, 8(3), 894–904, 1993.
12. L. Vargas, V. H. Quintana, Dynamic Analysis of Voltage Collapse in Longitudinal Systems, *Electrical Power and Energy Systems*, 16(4), 1994.
13. R. A. Schlueter, I.-P. Hu, Types of Voltage Instability and the Associated Modeling for Transient/Mid-Term Stability Simulation, *Journal of Electric Power System Research*, 29, 131–145, 1994.
14. A. Chakrabarti, D. P. Kothari, A. K. Mukhopadhyay, *Performance, Operation and Control of EHV Power Transmission System*, 1st ed., Wheeler Publication, New Delhi, India, 1995.
15. M. A. Pai, P. W. Sauer, B. C. Lesieutre, Static and Dynamic Nonlinear Loads and Structural Stability in Power Systems, *IEEE Proceedings, Special Issue on Nonlinear Phenomena in Power Systems*, 83(1)1, 1562–1572, 1995.
16. A. Chakrabarti, C. K. Chanda, G. Panda, Effects of Load Composition on Dynamic and Steady State Voltage Stability in a Longitudinal AC Power Transmission System, *Journal of AMSE – Association for the Advancement of Modeling and Simulation Techniques in Enterprises (France)*, 74(7), 8, 2001.
17. A. Mohamed, G. B. Jamson, A New Clustering Technique for Power System Voltage Stability Analysis, *Electrical Machines and Power Systems*, 23(4), 1995.

18. T. Lie, R. A. Schlueter, P. A. Rusche, Method of Identifying Weak Transmission Network Stability Boundaries, *IEEE Transactions on Power Systems*, 8, 293–301, 1993.

19. M. Stubbe, A. Bihain, J. Deuse, Simulation of Voltage Collapse, *Electrical Power and Energy Systems*, 15(4), 1993.

20. Y. Y. Hong, C. H. Gau, Voltage Stability Indicator for Identification of the Weakest Bus/Area in Power Systems, *IEE Proceedings of Generation, Transmission and Distribution*, 141(4), 305–309, 1994.

21. A. C. Zambroni, V. H. Quintana, New Technique of Network Partitioning for Voltage Collapse Margin Calculations, *IEE Proceedings of Generation, Transmission and Distribution*, 141(6), 630–636, 1994.

22. B. Lee, V. Ajjarapu, A Piecewise Global Small-Disturbance Voltage Stability Analysis of Structure Preserving Power System Models, *IEEE Transactions on Power Systems*, 10(4), 1963–1971, 1995.

23. A. Teshome, E. Esiyok, Distance to Voltage Collapse through Second Order Eigen-Value Sensitivity Technique, *Electrical Power and Energy Systems*, 17(6), 1995.

24. CIGRE Task Force 38-02-10, *Modeling of Voltage Collapse Including Dynamic Phenomena*, CIGRE, Paris, 1993.

25. CIGRE Task Force 38-02-11, *Indices Predicting Voltage Collapse Including Dynamic Phenomena*, CIGRE, Paris, July 1994.

26. A. Chakrabarti, A. K. Mukhopadhyay, Dynamic Damping of Power Swing and Improvement of Transient Voltage Stability with a Microprocessor Based SVC at Load Bus of a Longitudinal Power Supply System, *Journal of Electrical Engineering*, 82, 1–6, 2001.

27. A. Chakrabarti, A. K. Mukhopadhyay, Voltage Stability of Longitudinal Power System, *Journal of Electrical Engineering*, 71(Pt. El-3), 146, 1990.

28. G. K. Morison, B. Gao, P. Kundur, Voltage Stability Analysis Using Static and Dynamic Approaches, *IEEE Transactions on Power Systems*, PWRS-8(3), 1159–1171, 1993.

29. P. Kundur, G. K. Morison, B. Gao, Practical Considerations in Voltage Stability Assessment, *Electrical Power and Energy Systems*, 15(4), 1993.

30. J. Bian, P. Rastgoufard, Power System Voltage Stability and Security Assessment, *Electric Power System Research*, 30, 197–201, 1994.

31. C. W. Taylor, *Power System Voltage Stability*, EPRI Power System Engineers Series, McGraw-Hill, New York, 1994.

32. P. Rousseaux, T. Van Cutsen, Fast Small Disturbance Analysis of Long Term Voltage Stability, in *12th PSCC Proceedings*, Dresden, 1996, pp. 295–302.

33. R. R. Zalapa, B. J. Cory, Reactive Reserve Management, *IEE Proceedings of Generation, Transmission and Distribution*, 142(1), 17–23, 1995.

34. T. Van Cutsen, C. D. Vournas, Voltage Stability Analysis in Transient and Mid-Term Time Scales, *IEEE Transactions on Power Systems*, 11(1), 146–154, 1996.

35. R. Yokoyama, T. Nimure, Y. Nakamishi, A Coordinated Control of Voltage and Reactive Power Heuristic Modeling and Approximated Reasoning, *IEEE Transactions on Power Systems*, 8(2), 1993.

36. P. W. Sauer, M. A. Pai, Power System Steady State Stability and the Load Flow Jacobian, *IEEE Transactions on Power Systems*, 5(4), 1374–1383, 1990.

37. A. Chakrabarti, P. Sen, An Analytical Investigation of Voltage Stability of an EHV Transmission Network Based on Load Flow Analysis, *Journal of the Institution of Engineers (India)*, 76(Pt. EL), 1995.

38. T. J. Overbye, R. P. Klump, Effective Calculations of Power System Low-Voltage Solutions, *IEEE Transactions on Power Systems*, 11(1), 75–82, 1996.
39. T. J. Overbye, Computation of a Practical Method to Restore Power Flow Solvability, *IEEE Transactions on Power Systems*, 10(1), 280–287, 1995.
40. C. Canizares, W. Rosehart, A. Berizzi, C. Bovo, Comparison of Voltage Security Constrained Optimal Power Flow Technique, in *IEEE/PES Summer Meeting*, Vancouver, BC, vol. 3, pp. 1680–1685, July 2001.
41. V. Ajjarapu, B. Lee, Bifurcation Theory and Its Application to Nonlinear Dynamical Phenomena in an Electrical Power System, *IEEE Transactions on Power Systems*, 7(1), 424–431, 1992.
42. C. A. Canizares, On Bifurcations, Voltage Collapse and Load Modeling, *IEEE Transactions on Power Systems*, 10(1), 512–522, 1995.
43. CIGRE Task Force 38-02-12, *Criteria and Counter Measures for Voltage Collapse*, CIGRE, Paris, 1994.
44. M. H. Haque, Use of V-I Characteristics as a Tool to Assess Static Voltage Stability Limit of a Power System, *IEE Proceedings of Generation, Transmission and Distribution*, 151(1), 1–7, 2004.
45. M. Begovic, A. Phadke, Control of Voltage Stability Using Sensitivity Analysis, *IEEE Transactions on Power Systems*, 7(1), 114–123, 1992.
46. T. Van Cutsen, An Approach to Corrective Control of Voltage Instability Using Simulation and Sensitivity, *IEEE Transactions on Power Systems*, 10(2), 616–622, 1995.
47. I. Musirin, T. K. Abdul Rahman, Estimating Maximum Loadability for Weak Bus Identification Using FVSI, *IEEE Power Engineering Review*, 50–52, 2002.
48. M. H. Haque, Determination of Steady-State Voltage Stability Limit Using P-Q Curve, *IEEE Power Engineering Review*, 71–72, 2002.
49. B. Gao, G. K. Morison, P. Kundur, Voltage Stability Evaluation Using Modal Analysis, *IEEE Transactions on Power Systems*, 7(4), 1529–1542, 1992.
50. Y.-H. Hong, C.-T. Pan, W.-W. Lei, Fast Calculation of a Voltage Stability Index of Power System, *IEEE Transactions on Power Systems*, 12(4), 1555–1560, 1997.
51. M. Moghavvemi, F. M. Omar, Technique for Contingency Monitoring and Voltage Collapse Prediction, *IEE Proceedings of Generation, Transmission and Distribution*, 54(6), 634–640, 1998.
52. M. Moghavvemi, O. Faruque, Real Time Contingency Evaluation and Ranking Technique, *IEE Proceedings of Generation, Transmission and Distribution*, 145(5), 517–523, 1998.
53. M. Turan, S. B. Demircioglu, M. A. Yalcin, Voltage Stability Evaluation by Using Maximum Power Transfer Phasor Diagram, *Journal of Applied Science*, 6(1)3, 2809–2812, 2006.
54. B. Gao, G. K. Morison, P. Kundur, Voltage Stability Evaluation Using Modal Analysis, *IEEE Transactions on Power Systems*, 7(4), 1529–1542, 1992.
55. L. J. Cai, I. Erlich, Power System Static Voltage Stability Analysis Considering All Active and Reactive Power Controls – Singular Value Approach, in *IEEE Conference – Power Tech*, 2007, pp. 660–667.
56. C. A. Canizaris, A. C. Z. Souza, V. H. Quintana, Comparison of Performance Indices for Detection of Proximity to Voltage Collapse, *IEEE Transactions on Power Systems*, 11(3), 1441–1450, 1996.
57. B. D. Thukaram, Optimal Reactive Power Dispatch Algorithm for Voltage Stability Improvement, *Electric Power and Energy Systems*, 18(7), 1996.

58. M. E. Aggoune, Power System Security Assessment Using Artificial Neural Networks, PhD thesis, University of Washington, 1988.
59. H. Mori, Y. Tamaru, An Artificial Neural-Net Based Technique for Power System Dynamic Stability with Kohonen Model, *IEEE Transactions on Power Systems*, 7(4), 856–865, 1992.
60. D. Srinivasan, Power System Security Assessment and Enhancement Using Artificial Neural Networks, in *Proceedings of International Conference on Energy Management and Power Delivery IEEE-EMPD2*, 1998, pp. 582–587.
61. D. Neibur, A. J. Germond, Power System Security Assessment Using the Kohonen Neural Network Classifier, *IEEE Transactions on Power Systems*, 7(4), 865–872, 1992.
62. V. Brandwain, Severity Indices for Contingency Screening in Dynamic Security Assessment, *IEEE Transactions on Power Systems*, 12(4), 1136–1142, 1997.
63. C. Dingguo, R. R. Mohler, Neural Network Based Load Modeling and Its Use in Voltage Stability Analyses, *IEEE Transactions on Control Systems Technology*, 11(4), 460–470, 2003.
64. A. A. El-Keib, X. Ma, Application of Artificial Neural Networks in Voltage Stability Assessment, *IEEE Transactions on Power Systems*, 10(4), 1890–1896, 1995.
65. D. Paul, Md. Sadiq Iqba, M. I. Hussain, A. H. M. Shahriar Parvez, Voltage Stability Assessment Using Neural Network in the Deregulated Market Environment, *International Journal of Computer and Electrical Engineering*, 2(4), 747–752, 2010.
66. S. Kamalasadan, D. Thukaram, A. K. Srivastava, A New Intelligent Algorithm for Online Voltage Stability Assessment and Monitoring, *International Journal of Electrical Power and Energy Systems*, 31(2–3), 100–110, 2009.
67. K. Yabe, J. Koda, N. W. Miller, Conceptual Designs of AI-Based Systems for Local Prediction of Voltage Collapse, *IEEE Transactions on Power Systems*, 11(1), 137–145, 1996.
68. H. B. Wan, A. O. Ekwue, Artificial Neural Network Based Contingency Ranking Method for Voltage Collapse, *International Journal of Electrical Power and Energy Systems*, 22(5), 349–354, 2000.
69. B. Suthar, R. Balasubramanian, Application of an ANN Based Voltage Stability Assessment Tool to Restructured Power System, in *Bulk Power System Dynamics and Control-VII, Revitilizing Operational Reliability*, August 2007, pp. 1–8.
70. S. Chakrabarty, B. Jayasurya, Multicontingency Voltage Stability Monitoring of a Power System Using an Adaptive Radial Basis Function Network, *International Journal of Electrical Power and Energy Systems*, 22(5), 349–354, 2007.
71. N. H. Hingorani, Flexible AC Transmission System, *IEEE Spectrum*, 40–45, 1993.
72. C. W. Brice III, Voltage Drop Calculations and Power Flow Studies for Rural Electric Distribution Lines, *IEEE Transactions on Industrial Application*, 28(4), 1992.
73. I. B. Viswanathan, Effect of Series Compensation on Voltage Instability of EHV Long Lines, *Electrical Power System Research Journal*, 6, 185–191, 1983.
74. C. W. Taylor, *Power System Voltage Stability and Control*, McGraw-Hill, New York, 1993.
75. M. D. Cox, A. Mirbod, A New Static Var Compensator for an Arc Furnace, *IEEE Transactions on Power Systems*, 1(3), 110–119, 1986.
76. E. R. Johnson, P. S. Hasler, R. J. Moran, C. H. Titus, Static High-Speed VAR Control for Arc Furnace Flicker Reduction, in *Proceedings of America Power Conference*, 1972, pp. 1097–1105.

77. G. D. Breuer, H. M. Rustebakke, R. A. Gibley, H. Jirar, The Use of Series Capacitors to Obtain Maximum EHV Transability, *IEEE Transactions on Power Systems*, 11(1), 137–145, 1996.

78. M. O. Hassan, S. J. Cheng, Z. A. Zakaria, Steady-State Modeling of SVC and TCSC for Power Flow Analysis, in *Proceedings of the International Multi Conference of Engineers and Computer Scientists*, 2009, vol. II.

79. N. G. Hingorani, L. Gyugyi, *Understanding of FACTS: Concepts and Technology of Flexible AC Transmission Systems*, IEEE Press, New York, 1999.

80. E. Acha, *FACTS: Modeling and Simulation in Power Networks*, Wiley, New York, 2004.

81. A. Edris, FACTS Technology Development: An Update, *IEEE Engineering Review*, 20(3), 4–9, 2000.

82. R. M. Mathur, R. K. Varma, *Thyristor Based FACTS Controllers for Electrical Transmission Systems*, Wiley, New York, 2002.

83. M. H. Haque, Determination of Steady State Voltage Stability Limit of a Power System in the Presence of SVC, in *Power Tech Proceedings*, 2001.

84. W. Breuer, D. Povh, D. Retzmann, E. Teltsch, X. Lei, Role of HVDC and FACTS in Future Power System, in *CEPSI*, Shanghai, 2004.

85. A. Gustafasson, M. Jeroense, Energy Efficiency with HVDC Technology, in *ABB Power System*, Raleigh, NC.

86. S. Ge, T. Chung, Optimal Active Power Flow Incorporating Power Flow Control Needs in Flexible AC Transmission Systems, *IEEE Transactions on Power Systems*, 14(2), 738–744, 1999.

87. Y. Xu, H. Chen, FACTS-Based Power Flow Control in Interconnected Power Systems, *IEEE Transactions on Power Systems*, 5(1), 257–262, 2000.

88. L. Gyugyi, T. Rietman, A. Edris, The UPFC Power Flow Controller: A New Approach to Power Transmission Control, *IEEE Transactions on Power Delivery*, 10(2), 1085–1092, 1995.

89. X. Zhang, E. Handschin, Advanced Implementation of UPFC in a Nonlinear Interior-Point OPF, *IEE Proceedings Generation Transmission Distribution*, 148(5), 489–496, 2001.

90. B. Venkatesh, M. K. George, H. B. Gooi, Fuzzy OPF Incorporating UPFC, *IEE Proceedings Generation Transmission Distribution*, 151(5), 625–629, 2004.

91. G. Taranto, L. Pinto, M. Pereita, Presentation of FACTS Devices in Power System Economic Dispatch, *IEEE Transactions on Power Systems*, 7(2), 572–576, 1992.

92. H. M. Haque, Power Flow Control and Voltage Stability Limit: Regulating Transformer versus UPFC, *IEE Proceedings Generation Transmission Distribution*, 151(3), 299–304, 2004.

93. M. Z. Al-Sadek, M. M. Dessouky, G. A. Mahmoud, W. I. Rashed, Enhancement of Steady State Voltage Stability by Static VAR Compensators, *Electric Power Systems Research*, 43(2), 179–185, 1997.

94. K. Visakha, D. Thukaram, L. Jenkins, Applications of UPFC for System Security Improvement under Normal and Network Contingencies, *Electric Power Systems Research*, 70(1), 46–55, 2004.

95. A. Sannino, J. Svensson, T. Larsson, Power Electronic Solutions to Power Quality Problems, *Electric Power Systems Research*, 66(1), 71–82, 2003.

96. J. Sun, D. Czarkkowski, Z. Zabar, Voltage Flicker Mitigation Using PWM-Based Distribution STATCOM, in *IEEE Power Engineering Society Summer Meeting*, 2002.

97. M. Noroozian, G. Andersson, Power Flow Control by Use of Controllable Series Components, *IEEE Transactions on PWRD*, 8(3), 1420–1429, 1993.

98. H. F. Wang, F. J. Swift, A Unified Model for the Analysis of FACTS Devices in Damping Power System Oscillations, Part I: Single-Machine Infinite-Bus Power Systems, *IEEE Transactions on PWRD*, 12(2), 941–946, 1997.

99. M. Z. El-Sadek, M. Dessouky, G. A. Mahmoud, W. I. Rashed, Series Capacitors Combined with Static VAR Compensators for Enhancement of Steady-State Voltage Stabilities, *Electric Power System Research*, 44(3), 137–143, 1998.

100. M. H. Haque, Determination of Steady-State Voltage Stability Limit of a Power System in the Presence of SVC, in *IEEE Porto Power Tech Conference*, 2001.

101. C. W. Taylor, Static Var Compensator Models for Power Flow and Dynamic Performance Simulation, *IEEE Transactions on Power Systems*, 9(1), 229–239, 1994.

102. I. A. Enrmez, *Static Var Compensator*, Working Group 38, Task Force 2 on SVC, CIGRE, 1986.

103. S. Lefebvre, L. G. Lajoie, A Static Compensator Model for EMTP, *IEEE Transactions on Power Systems*, 7(2), 477–486, 1992.

104. Y. Mansour, W. Xu, C. Rinzin, SVC Placement Using Critical Modes of Voltage Instability, *IEEE Transactions on Power Systems*, 9(2), 757–763, 1994.

105. C. R. Fuerte-Esquivel, E. Acha, Newton-Raphson Algorithm for the Reliable Solution of Large Power Networks with Embedded FACTS Devices, *IEE Proceedings of Generation, Transmission and Distribution*, 143(5), 447–454, 1996.

106. C. R. Fuerte-Esquivel, E. Acha, H. A. Perez, A Thyristor Controlled Series Compensator Model for the Power Flow Solution of Practical Power Networks, *IEEE Transactions on Power Systems*, 15(1), 58–64, 2000.

107. T. V. Trijillo, C. R. Fuerte-Esquivel, J. H. T. Hernandez, Advance Three Phase Static Var Compensator Models for Power Flow Analysis, *IEE Proceedings of Generation, Transmission and Distribution*, 150(1), 119–127, 2003.

108. S. Arabi, P. Kundur, A Versatile FACTS Device Model for Power Flow and Stability Simulations, *IEEE Transactions on Power Systems*, 11(4), 1944–1950, 1996.

109. M. H. Haque, Maximum Power Transfer Capability within the Voltage Stability Limit of Series and Shunt Compensation Schemes for AC Transmission System, *Electric Power System Research*, 24, 227–235, 1992.

110. S. Dey, C. K. Chanda, A. Chakrabarti, Development of Global Voltage Security Indicator and Role of SVC on It in Longitudinal Power Supply System, *Electric Power System Research*, 68, 1–9, 2004.

111. M. Noroozian, I. Hiskens, M. Ghandhari, A Robust Control Strategy for Shunt and Series Reactive Compensators to Damp Electromechanical Oscillations, *IEEE Transactions on Power Systems*, 16(4), 1944–1950, 2001.

112. C. R. Fuerte-Esquivel, E. Acha, S. G. Tan, J. J. Rico, Efficient Object Oriented Power Systems Software for the Analysis of Large Scale Networks Containing FACTS-Controlled Branches, *IEEE Transactions on Power Systems*, 13(2), 464–472, 1998.

113. Z. Jinhua, W. Mingxin, L. Liangjun, Reliability Studies for Three Gorges DC System, *Proceedings of POWERCON*, 1, 456–459, 1998.

114. D. A. Waterworth, C. P. Arnold, N. R. Watson, Reliability Assessment Technique for HVDC Systems, *IPENZ Transactions*, 25(1), 1998.

115. D. Jovcic, N. Pahalawaththa, M. Zavahir, Inverter Controller for HVDC Systems Connected to Weak AC Systems, *IEE Proceedings of Generation, Transmission and Distribution*, 146(3), 235–240, 1999.

116. G.-J. Li, S. Ruan, L. Peng, Y. Sun, X. Li, A Novel Nonlinear Control for Stability Improvement in HVDC Light System, in *Proceedings of IEEE Conference*, 2005.

117. M. Bahrman, HVDC Transmission, in *IEEE PSCE*, Atlanta, GA, November 2006.
118. W. Breuer, D. Povh, E. Teltsch, Trends for Future HVDC Application, in *16th Conference of the Electric Power Supply Industry*, 2006.
119. H. F. Latorre, M. Ghandhari, Improvement of Voltage Stability by Using VSC-HDC, in *IEEE T&D ASIA – SIEF*, Seoul, Korea, October 2009.
120. L. C. Azimoh, K. Folly, S. P. Chowdhury, S. Chowdhury, Investigation of Voltage and Transient Stability of HVAC Network in Hybrid with VSC-HVDC and HVDC Link, in *Proceedings of UPEC*, 2010.
121. Y. Pipelzadeh, B. Chaudhuri, T. C. Green, Stability Improvement through HVDC Upgrade in the Australian Equivalent System, in *Proceedings of UPEC*, 2010.
122. A. L'Abbate, G. Fulli, Modeling and Application of VSC-HVDC in the European Transmission System, *International Journal of Innovations in Energy Systems and Power*, 5(1), 2010.
123. N. Nayak, S. K. Routray, P. K. Rout, Improvement of Transient Stability of VSC HVDC System with Particle Swarm Optimization Based PI Controller, *International Journal of Power System Operation and Energy Management (IJPSOEM)*, I, 81–89, 2011.
124. J. C. Kaltenbach, J. Peschon, E. H. Gehrig, A Mathematical Optimization Technique for the Expansion of Electric Power Transmission Systems, *IEEE Transactions on Power Apparatus and Systems*, PAS-89(1), 113–119, 1970.
125. Y. N. Yu, K. Vongsuriyaber, L. N. Wedman, Application of an Optimal Control Theory to a Power System, *IEEE Transactions on Power Apparatus and Systems*, PAS-89(1), 55–62, 1970.
126. A. M. Sasson, H. M. Merrill, Some Application on Optimization Techniques to Power System Problem, *Proceedings of IEEE*, 62(5), 1974.
127. J. Nanda, D. P. Kothari, S. C. Srivastava, A New Optimal Dispatch Algorithm Using Fletcher's QP Method, *Proceedings of IEE*, 136(3), 153–161, 1989.
128. S. K. Agarwal, I. J. Nagrath, Optimal Scheduling of Hydro-Thermal System, *Proceedings of IEE*, 169(2), 199–207, 1972.
129. A. K. Ayub, A. D. Patton, Optimal Thermal Generating Unit Commitment, *IEEE Transactions on Power Apparatus and Systems*, 90(1), 1752–1761, 1971.
130. J. F. Dopazo, An Optimizing Technique for Real and Reactive Power Allocation, in *Proceedings of IEEE*, November 1967.
131. D. P. Kothari, Optimal Hydrothermal Scheduling and Unit Commitment, PhD thesis, BITS, Pilani, 1975.
132. D. P. Kothari, I. J. Nagrath, Security Constrained Economic Thermal Generating Unit Commitment, *Journal of Institute of Engineers*, 59, 156–163, 1978.
133. H. W. Dommel, W. F. Trinney, Optimal Power Flow Solution, *IEEE Transactions on Power Apparatus and Systems*, 87(1),1866–1876, 1968.
134. G. L. Kusic, *Computer Aided Power System Analysis*, Prentice-Hall, Englewood Cliffs, NJ, 1986.
135. R. Billinton, *Power System Reliability Evaluation*, Gordon and Breach, New York, 1970.
136. R. Billinton, R. J. Ringlee, A. J. Wood, *Power System Reliability Calculations*, MIT Press, Boston, MA, 1973.
137. J. Kennedy, R. Eberhart, Particle Swarm Optimization, in *IEEE International Conference on Neural Network*, Australia, 1995, pp. 1942–1948.
138. J. B. Park, K. S. Lee, Particle Swarm Optimization for Economic Dispatch with Non-Smooth Cost Function, *IEEE Transactions on Power Systems*, 20(1), 34–42, 2005.

139. D. M. Vinod Kumar, C. Venkaiah, Swarm Intelligent Based Security Constrained Congestion Management Using SSSC, in *IEEE Power and Energy Engineering*, APPEC, 2009, pp. 1–6.

140. S. Dutta, S. P. Singh, Optimal Rescheduling of Generators for Congestion Management Based on Particle Swarm Optimization, *IEEE Transactions on Power Systems*, 23(4), 1560–1569, 2008.

141. F. He, Y. Wang, K. W. Chan, Y. Zhang, Optimal Load Shedding Strategy Based on Particle Swarm Optimization, in *8th International Conference on Advances in Power System Control Operation and Management*, 2009.

142. N. Mo, Z. Y. Zou, K. W. Chan, T. Y. G. Pong, Transient Stability Constrained Optimal Power Flow Using Particle Swarm Optimization, *IEE Proceedings of Generation, Transmission and Distribution*, 1(3), 476–483, 2007.

143. M. Saravanan, S. M. R. Slochanal, P. Venkatesh, J. P. S. Abraham, Application of Particle Swarm Optimization Technique for Optimal Location of FACTS Devices Considering Cost of Installation and System Loadability, *Electrical Power System Research*, 77, 276–283, 2007.

144. M. A. Abido, Multi-Objective Particle Swarm Optimization for Environmental/ Economic Dispatch Problem, *Electric Power Systems Research*, 79(7), 1105–1113, 2009.

145. L. L. Lai, J. T. Yokoyama, M. Zhao, Improved Genetic Algorithms for Optimal Power Flow under Both Normal and Contingent Operation States, *International Journal of Electrical Power and Energy Systems*, 19(5), 287–292, 1997.

146. A. Chuang, W. Felix, An Extensible Genetic Algorithm Framework for Problem Solving in a Common Environment, *IEEE Transactions on Power Systems*, 15(1), 269–275, 2000.

147. M. S. Osama, M. A. Abo-Sinna, A. A. Mousa, A Solution of Optimal Power Flow by Genetic Algorithm, *International Journal of Applied Mathematics and Computation*, 155(2), 391–405, 2004.

148. M. Todorovski, D. Rajičić, An Initialization Procedure in Solving Optimal Power Flow by Genetic Algorithm, *IEEE Transactions on Power Systems*, 21(2), 480–487, 2006.

149. Y. Wei, Y. Juan, C. Y. David, B. Kalu, A New Optimal Reactive Power Flow Model in Rectangular Form and Its Solution by Predictor Corrector Primal Dual Interior Point Method, *IEEE Transactions on Power Systems*, 21(1), 61–67, 2006.

150. A. J. Rabih, Optimal Power Flow Using an Extended Conic Quadratic Formulation, *IEEE Transactions on Power Systems*, 23(3), 1000–1008, 2008.

151. M. R. Rashidi, M. E. Hawary, Applications of Computational Intelligence Techniques for Solving the Revived Optimal Power Flow Problem, *Electric Power Systems Research*, 79(4), 694–702, 2009.

152. L. J. Cai, I. Erlich, G. Stamtsis, Optimal Choice and Allocation of FACTS Devices in Deregulated Electricity Market Using Genetic Algorithms, in *Power System Conference and Expositions*, 2004, pp. 201–207.

153. M. S. Osman, M. A. Abo-Sinna, A. A. Mousa, A Solution to the Optimal Power Flow Using Genetic Algorithm, *International Journal of Applied Mathematics and Computation*, 155(2), 391–405, 2004.

154. T. Mirko, R. Dragoslav, An Initialization Procedure in Solving Optimal Power Flow by Genetic Algorithm, *IEEE Transactions on Power Systems*, 21(2), 480–487, 2006.

155. R. Storn, K. Price, *Differential Evolution – A Simple and Efficient Adaptive Scheme for Global Optimization over Continuous Spaces*, Technical Report, TR-95-012, Berkeley, CA, 1995.
156. K. P. Wong, Z. Y. Dong, Differential Evolution, an Alternative Approach to Evolutionary Algorithm, in *ISAP*, 2005, pp. 73–83.
157. H. R. Cai, C. Y. Chung, K. P. Wong, Application of Differential Evolution Algorithm for Transient Stability Constrained Optimal Power Flow, *IEEE Transactions on Power Systems*, 23(2), 719–728, 2008.
158. L. D. S. Coelho, V. C. Mariani, Combining of Chaotic Differential Evolution and Quadratic Programming for Economic Dispatch Optimization with Valve-Point Effect, *IEEE Transactions on Power Systems*, 21(2), 989–996, 2006.
159. S. L. Cheng, C. Hwang, Optimal Approximation of Linear Systems by a Differential Evolution Algorithm, *IEEE Transactions on Systems, Man, and Cybernetics – Part A: Systems and Humans*, 31(6), 698–707, 2001.
160. R. Gamperie, S. Muller, A Parameter Study for Differential Evolution, *Advances in Intelligent Systems, Fuzzy Systems, Evolutionary Computation*, WSEAS Press, Athens, Greece, 2002, pp. 293–298.
161. S. Rahnamayan, H. R. Tizhoosh, M. M. A. Salma, Opposition Based Differential Evolution, *IEEE Transactions on Evolutionary Computation*, 12(1), 64–79, 2008.
162. M. M. Metwally, A. Emary, F. M. Bendary, M. I. Mosaad, Optimal Power Flow Using Evolutionary Programming Techniques, *IEEE Transactions on Power Systems*, 23(5), 260–264, 2008.
163. G. V. John, Y. Lee, Quantum Inspired Evolutionary Algorithm for Real and Reactive Power Dispatch, *IEEE Transactions on Power Systems*, 23(4), 1627–1636, 2008.
164. S. K. Bath, J. S. Dhillon, D. P. Kothari, Fuzzy Satisfying Multi-Objective Generation Scheduling by Weightages Pattern Search Method, *International Journal of Electrical Power System and Research*, 69, 311–320, 2004.
165. V. C. Ramesh, Optimal Power Flow with Fuzzy Emissions Constraints, *International Journal of Electrical Machines and Power System*, 25(1), 897–906, 1997.
166. D. Hur, B. Kim, Application of Distributed Optimal Power Flow to Power System Security Assessment, *International Journal of Electric Power Components and Systems*, 31(1), 71–80, 2003.
167. J. S. Dhillon, S. C. Parti, D. P. Kothari, Fuzzy Decision Making in Multi-Objective Long-Term Scheduling of Hydrothermal System, *International Journal of Electrical Power System and Research*, 23(1), 19–29, 2001.
168. Y. S. Brar, J. S. Dhillon, D. P. Kothari, Multi-Objective Load Dispatch by Fuzzy Logic Based Searching Weightages Pattern, *International Journal of Electrical Power System and Research*, 63, 149–160, 2002.
169. J. S. Dhillon, S. C. Parti, D. P. Kothari, Fuzzy Decision Making in Stochastic Multi-Objective Short-Term Hydrothermal Scheduling, *IEE Proceedings of Generation, Transmission and Distribution*, 149(2), 191–200, 2002.
170. M. Davison, C. L. Anderson, B. Marcus, K. Anderson, Development of Hybrid Model for Electrical Power Spot Price, *IEEE Transactions on Power Systems*, 17(2), 2002.
171. K. Xie, Y. H. Song, J. Stonham, E. Yu, G. Lie, Decomposition Model and Interior Point Methods for Optimal Spot Pricing of Electricity in Deregulation Environments, *IEEE Transactions on Power Systems*, 15(1), 39–50, 2000.

172. L. Goel, P. A. Viswanath, P. Wang, Reliability Evaluation of Hybrid Power Markets Based on Multiple Transaction Matrix and Optimal Transaction Curtailment, *IEE Proceedings of Generation, Transmission and Distribution*, 145(1), 66–70, 2007.
173. IEEE Committee Report, *Economy-Security Functions in Power System Operations*, IEEE Special Publication 75 CHO 969.6 PWR, New York, 1975.
174. T. W. Berrie, *Power System Economics*, IEE, London, 1983.
175. T. W. Berrie, *Electricity, Economics and Planning*, IEE, London, 1992.
176. I. J. Nagrath, D. P. Kothari, Optimal Stochastic Scheduling of Cascaded Hydrothermal System, *Journal of Institute of Engineers*, 56, 264–270, 1976.
177. J. Paschen, Optimal Control of Reactive Power Flow, *IEEE Transactions on Power Apparatus and Systems*, 87, 40–48, 1968.
178. F. Wu, A Two Stage Approach to Solving Optimal Power Flow, in *Proceedings of PICA Conference*, 1979, pp. 126–136.
179. D. P. Kothari, Some Aspects of Optimal Maintenance Scheduling of Generating Units, *Journal of Institute of Engineers*, 66, 41–50, 1985.
180. J. S. Dhillon, S. C. Parti, D. P. Kothari, Multi-Objective Optimal Thermal Power Dispatch, *International Journal of Electrical Power System and Research*, 16(6), 383–389, 1994.
181. J. A. Momoh, R. J. Koessler, M. S. Bond, B. Stott, D. Sun, A. Papalexopoulos, P. Ristanovic, Challenges to Optimal Power Flow, in *IEEE/PES Winter Meeting*, January 1996.
182. R. Bacher, H. P. Van Meeteren, Real Time Optimal Power Flow in Automatic Generation Control, *IEEE Transactions on Power Systems*, 3(2), 1518–1529, 1988.
183. G. Anastasios, N. B. Pandel, V. Petridis, Optimal Power Flow by Enhanced Genetic Algorithm, *IEEE Transactions on Power Systems*, 17(2), 229–236, 2002.
184. R. Suresh, N. Kumarappan, Genetic Algorithm Based Reactive Power Optimization under Deregulation, in *IET-UK International Conference on Information and Communication Technology in Electrical Sciences*, December 2007, pp. 150–157.
185. L. Nepomuceno, A. Santos, Equivalent Optimization Model for Loss Minimization: A Suitable Analysis Approach, *IEEE Transactions on Power Systems*, 12(4), 1403–1412, 1997.
186. J. A. Momoh, M. E. El-hawary, R. Adapa, A Review of Selected Optimal Power Flow Literature to 1993, *IEEE Transactions on Power Systems*, 14(Pt. I–II), 96–111, 1999.
187. R. D. Christie, B. F. Wollenberg, I. Wangensteen, Transmission Management in the Deregulated Environment, *Proceedings of IEEE*, 88, 170–195, 2000.
188. O. Alasac, B. Stott, Optimal Power Flow with Steady State Security, *IEEE Transactions on Power Apparatus and Systems*, PAS-93, 745–751, 1974.
189. J. A. Momoh, A Generalized Quadratic-Based Model for Optimal Power Flow, *IEEE Transactions on Systems, Man, and Cybernetics*, SMC-16, 1986.
190. G. F. Reid, L. Hasdorf, Economic Dispatch Using Quadratic Programming, *IEEE Transactions on Power Apparatus and Systems*, PAS-92, 2015–2023, 1973.
191. D. E. Goldberg, *Genetic Algorithm in Search, Optimisation and Machine Learning*, 3rd impression, Pearson Education, India.
192. F. C. Schweppe, M. C. Caramanis, R. E. Bohn, Optimal Spot Pricing: Practice and Theory, *IEEE Transactions on Power Apparatus and Systems*, PAS-101(9), 1982.
193. F. C. Schweppe, M. C. Caramanis, R. D. Tabors, R. E. Bohn, *Spot Pricing of Electricity*, Kluwer Academic Publishers, Dordrecht, Netherlands, 1988.

194. F. Yong, F. Zuyi, Different Models and Properties on LMP Calculations, in *Power Engineering Society, General Meeting, IEEE*, 2006.
195. A. A. El-Keib, X. Ma, Calculating Short-Run Marginal Costs of Active and Reactive Power Production, *IEEE Transactions on Power Systems*, 12(2), 559–565, 1997.
196. M. Aganagic, K. H. Abdul-Rahman, J. G. Waight, Spot Pricing of Capacities for Generation and Transmission of Reserve in an Extended Poolco Model, *IEEE Transactions on Power Systems*, 13(3), 1128–1135, 1998.
197. J. M. Arroyo, A. J. Conejo, Optimal Response of a Thermal Unit to an Electricity Spot Market, *IEEE Transactions on Power Systems*, 15(3), 1098–1104, 2000.
198. A. Urkmez, N. Cetinkaya, Determining Spot Price and Economic Dispatch in Deregulated Power Systems, *Mathematical and Computational Applications*, 15(1), 25–33, 2010.
199. S. Parnandi, K. Schoder, A. Feliachi, Power Market Analysis Tools for Congestion Management, *Journal of Electrical System*, 4(3), 2008.
200. E. L. Silva, J. J. Hedgecock, J. Mello, J. Luz, Practical Cost Based Approach for the Voltage Ancillary Services, *IEEE Transactions on Power Systems*, 16(4), 806–812, 2001.
201. A. Kumar, S. C. Srivastava, AC Power Transfer Distribution Factors for Allocating Power Transactions in a Deregulated Market, *IEEE Power Engineering Review*, 42–43, 2002.
202. M. Davison, C. L. Anderson, B. Marcus, K. Anderson, Development of a Hybrid Model for Electrical Power Spot Prices, *IEEE Transactions on Power Systems*, 17(2), 257–264, 2002.
203. K. Xie, Y. H. Song, J. Stonham, E. Yu, G. Li, Decomposition Model and Interior Point Methods for Optimal Spot Pricing of Electricity in Deregulation Environments, *IEEE Transactions on Power Systems*, 15(2), 39–50, 2000.
204. New York Independent System Operator, 2000, www.nyiso.com.
205. California Independent System Operator, 2000, www.caliso.com.
206. S. Deng, Financial Methods in Competitive Electricity Markets, PhD dissertation, Department of Industrial Engineering, University of California, Berkeley, 1999.
207. P. J. Joskow, Electricity Sector in Transition, *Energy Journal*, 9(2), 25–52, 1998.
208. B. Anderson, L. Bregman, Market Structure and the Price of Electricity, *Energy Journal*, 16(2), 97–109, 1995.
209. N. Dandachi, M. Rawlins, O. Alsac, M. Prais, B. Scott, OPF for Reactive Pricing Studies on NGC System, *IEEE Transactions on Power Systems*, 11(1), 42–50, 1996.
210. L. Cai, Y. Luo, G. Stamtsis, I. Erlich, Optimal Choice and Allocation of FACTS Devices in Deregulated Electricity Market Using Genetic Algorithms, *IEEE, PES, Power System Conference and Expositions*, 1, 201–207, 2004.
211. T. T. Lie, W. Deng, Optimal Flexible AC Transmission Systems Devices Allocation, *Electrical Power and Energy System*, 9(2), 125–134, 1997.
212. T. S. Chung, Y. Z. Li, A Hybrid GA Approach for OPF with Consideration of FACTS Devices, *IEEE Power Engineering Review*, 47–57, 2001.
213. W. L. Fang, H. W. Ngan, Optimizing Location of Unified Power Flow Controllers Using the Method of Augmented Lagrange Multipliers, *IEE Proceedings of Generation, Transmission and Distribution*, 146(2), 428–434, 1999.
214. F. D. Galiana, K. Almeida, M. Toussaint, J. Griffin, Assessment and Control of the Impact on FACTS Devices on Power System Performance, *IEEE Transactions on Power Systems*, 11(4), 42–50, 1996.

215. C. D. Voumas, Interruptible Load as a Competitor to Local Generation for Preventing Voltage Security, in *IEEE Power Engineering Society Winter Meeting,* 2001, vol. 1, pp. 236–240.

216. M. Shahidehpour, M. Alomoush, *Restructured Electrical Power Systems: Operation, Trading and Volatility,* Marcel Dekker, New York, 2001.

217. G. L. Doorman, Optimal System Security under Capacity Constrained Conditions, *IEEE Porto Power Tech Proceedings,* 2(6), 12–19, 2001.

218. I. Kopcak, L. C. P. Silva, J. S. Naturesa, Transmission System Congestion Management by Using Modal Participation Factors, in *IEEE Bologna Power Tech Conference,* Bologna, Italy, 1–6, 2003.

219. N. P. Padhy, Congestion Management under Deregulated Fuzzy Environment, in *IEEE International Conference on Electric Utility Deregulation, Restructuring and Power Technologies (DRPT),* April 2004, pp. 133–139.

220. Y. Peng, Y. Bing, S. Jiahua, Comparison Study of Spot Price under Transmission Congestion with Different Control Mechanism, in *IEEE Transmission and Distribution Conference and Exhibition: Asia and Pacific,* Dalian, China, 2005, pp. 1–5.

221. J. Ma, Y. H. Song, O. Lu, S. Mei, Framework for Dynamic Congestion Management in Open Power Markets, *IEE Proceedings of Generation, Transmission and Distribution,* 149(2), 157–164, 2002.

222. R. S. Fang, A. K. David, Transmission Congestion Management in an Electricity Market, *IEEE Transactions on Power Systems,* 14(3), 877–883, 1999.

223. E. M. Yap, M. A. Dabbagh, P. C. Thum, UPFC Controller in Mitigating Line Congestion for Cost-Efficient Power Delivery, in *Power Engineering Conference, IPEC,* 2005, pp. 1–6.

224. V. Z. Farahani, A. Kazmi, A. B. Majd, Congestion Management in Bilateral Based Power Market by FACTS Devices and Load Curtailments, in *IEEE Power India Conference,* 2006, pp. 1–6.

225. Z. Feng, V. Ajjarapu, A Comprehensive Approach for Preventive and Corrective Control to Mitigate Congestion Ensuring Voltage Stability, *IEEE Transactions on Power Systems,* 15(4), 791–797, 2000.

226. X. P. Zhang, L. Yao, A Vision of Electricity Network Congestion Management with FACTS and HVDC, in *IEEE International Conference on Electric Utility Deregulation, Restructuring and Power Technologies (DRPT),* April 2008, pp. 116–121.

227. V. P. Rajderkar, V. K. Chandrakar, Optimal Location of Thyristor Controlled Series Compensator for Congestion Management, *Journal of Institute of Engineers,* 90, 30–35, 2009.

228. L. Goel, P. A. Viswanath, P. Wang, Reliability Evaluation of Hybrid Power Markets Based on Multiple Transaction Matrix and Optimal Transaction Curtailment, *IET Proceedings of Generation, Transmission and Distribution,* 1(1), 67–71, 2007.

229. L. P. Hajdu, J. Peschon, D. S. Piercy, W. F. Tinney, Optimum Load Shedding Policy for Power Systems, *IEEE Transactions on Power Apparatus and Systems,* PAS-87(3), 1968.

230. J. Hazra, A. K. Sinha, Congestion Management Using Multi-Objective Particle Swarm Optimization, *IEEE Transactions on Power Systems,* 22(4), 1726–1734, 2007.

231. S. Dutta, S. P. Singh, Optimal Rescheduling of Generators for Congestion Management Based on Particle Swarm Optimization, *IEEE Transactions on Power Systems,* 23(4), 1560–1569, 2008.

232. A. K. David, Dispatch Methodologies for Open Access Transmission Systems, *IEEE Transactions on Power Systems*, 13(1), 46–53, 1998.

233. D. M. Vinod Kumar, C. Venkaiah, Swarm Intelligence Based Security Constrained Congestion Management Using SSSC, in *Power and Energy Engineering Conference, APPEEC, Asia-Pacific*, 2009.

234. E. Muneender, D. M. Vinod Kumar, Optimal Real and Reactive Power Dispatch for Zonal Congestion Management Problem for Multi-Congestion Case Using Adaptive Fuzzy PSO, in *TENCON*, 2009, pp. 1–7.

235. M. A. Rahim, I. Musirin, I. Z. Abidin, D. Joshi, Congestion Management Based Optimization Technique Using Bee Colony, in *4th International Power Engineering and Optimization Conference (PEOCO)*, Malaysia, June 2010, pp. 184–188.

236. S. Balaraman, N. Kamaraj, Congestion Management in Deregulated Power System Using Real Coded Genetic Algorithm, *International Journal of Engineering Science and Technology*, 11(2), 6681–6690, 2010.

237. S. Balaraman, N. Kamaraj, Application of Differential Evolution for Congestion Management in Power System, *Modern Applied Science, Canadian Center of Science and Education*, 4(8), 33–43, 2010.

238. K. Selvi, T. Meena, N. Ramaraj, A Generation Rescheduling Method to Alleviate Line Overloads Using PSO, *Journal of Institute of Engineers (India)*, 88, 10–14, 2007.

239. H. Yang, M. Lai, A Congestion Cost Allocating Method Based on Aumann-Shapley Value in Bilateral Model, in *IEEE, Power Engineering Society, General Meeting*, July 2003, vol. 2.

240. S. Raikar, M. Ilic, Assessment of Transmission Congestion for Major Electricity Markets in US, in *IEEE, Power Engineering Society, Summer Meeting*, July 2001, vol. 2, pp. 1452–1456.

241. L. Nepomuceno, A. Santos Jr., Equivalent Optimization Model for Loss Minimization: A Suitable Analysis Approach, *IEEE Transactions on Power Systems*, 12(4), 1403–1412, 1997.

242. G. Gross, S. Tao, A Physical-Flow-Based Approach to Allocating Transmission Losses in a Transaction Framework, *IEEE Transactions on Power Systems*, 15(2), 631–637, 2000.

243. Y.-C. Chang, C.-N. Lu, An Electricity Tracing Method with Application to Power Loss Allocation, *International Journal of Electrical Power Energy Systems*, 23(1), 13–17, 2001.

244. S. Abdelkar, Transmission Loss Allocation in a Deregulated Electrical Energy Market, *Electrical Power System Research*, 76, 962–967, 2006.

245. G. Strbac, D. Kirschen, S. Ahmed, Allocating Transmission System Usage on the Basis of Traceable Contribution of Generators and Loads to Flow, *IEEE Transactions on Power Systems*, 13(2), 527–534, 1998.

246. T. K. Hann, J. H. Kim, J. K. Park, Calculation of Transmission Loss Factor Considering Load Variation, in *IEEE Power Engineering Society Meeting*, July 2002, pp. 21–25.

247. K. Min, S.-H. Ha, S.-W. Lee, Y.-H. Moon, Transmission Loss Allocation Algorithm Using Path-Integral Based on Transaction Strategy, *IEEE Transactions on Power Systems*, 25(1), 195–205, 2010.

248. M. H. Sulaiman, M. W. Mustafa, O. Aliman, Transmission Loss and Load Flow Allocations via Genetic Algorithm Technique, in *IEEE Conference, TENCON*, 2009, pp. 1–7.

249. J. H. Teng, Power Flow and Loss Allocation for Deregulated Transmission Systems, *Electric Power Systems Research*, 27, 327–333, 2005.
250. S. Abdelkar, Efficient Computation Algorithm for Calculating Load Contributions to Line Flows and Losses, *IEE Proceedings of Generation, Transmission and Distribution*, 153(4), 391–398, 2006.
251. S. Hao, A Reactive Power Management Proposal for Transmission Operators, *IEEE Transactions on Power Systems*, 18(4), 1374–1381, 2003.
252. R. Suresh, N. Kumarappan, Genetic Algorithm Based Reactive Power Optimization under Deregulation, in *IET-UK International Conference on Information and Communication Technology in Electrical Science*, December 2007, pp. 150–155.
253. U.S. Department of Energy, *The Smart Grid: An Introduction*, U.S. Department of Energy Washington, DC, http://www.oe.energy.gov/1165.htm, 2008.
254. European Smart Grids Technology Platform Vision and Strategy for Europe's Electricity Networks of the Future, http://ec.europa.eu/research/energy/pdf/smartgrids_en.pdf.
255. K. Moslehi, R. Kumar, Smart Grid – A Reliability Perspective, in *IEEE PES Conference on Innovative Smart Grid Technologies*, January 19–20, 2010.
256. F. Li, W. Qiao, H. Sun, H. Wan, J. Wang, Y. Xia, Z. Xu, P. Zhang, Smart Transmission Grid: Vision and Framework, *IEEE Transactions on Smart Grid*, 1(2), 168–177, 2010.
257. Z. X. Song, C. Li-Qiang, M. You-Jie, Research on Smart Grid Technology, in *International Conference on Computer Application and System Modeling (ICCASM)*, 2010, pp. 599–603.
258. A. Moshari, G. R. Yousefi, A. Ebrahimi, S. Haghbin, Demand-Side Behavior in the Smart Grid, Environment, in *Innovative Smart Grid Technologies (ISGT)*, January 2010, pp. 1–7.
259. K. Y. Huang, H. C. Chin, Y. C. Huang, A Model Reference Adaptive Control Strategy for Interruptible Load Management, *IEEE Transactions on Power Systems*, 19(1), 683–689, 2004.
260. Y. Huang, A. Brocco, P. Kuonen, M. Courant, B. Hirsbrunner, Smart GRID: A Fully Decentralized Grid Scheduling Framework Supported by Swarm Intelligence, in *Seventh International Conference on Grid and Cooperative Computing*, October 2008, pp. 24–26.
261. M. M. Abdullah, B. Dwolatzky, Smart Demand-Side Energy Management Based on Cellular Technology – A Way towards Smart Grid Technologies in Africa and Low Budget Economies, in *IEEE AFRICON*, September 2009, pp. 23–25.
262. N. Ruiz, I. Cobelo, J. Oyarzabal, A Direct Load Control Model for Virtual Power Plant Management, *IEEE Transactions on Power Systems*, 24(2), 959–966, 2009.
263. J. Medina, N. Muller, I. Roytelman, Demand Response and Distribution Grid Operations: Opportunities and Challenges, *IEEE Transactions on Smart Grid*, 1(2), 2010.
264. M. A. S. Masoum, P. S. Moses, S. Deilami, Load Management in Smart Grids Considering Harmonic Distortion and Transformer Derating, in *Innovative Smart Grid Technologies (ISGT)*, January 2010, pp. 1–7.
265. N. Navid-Azarbaijani, M. H. Banakar, Realizing Load Reduction Functions by Aperiodic Switching of Load Groups, *IEEE Transactions on Power Systems*, 11(2), 721–727, 1996.

266. F. Rahimi, A. Ipakchi, Demand Response as a Market Resource under the Smart Grid Paradigm, *IEEE Transactions on Smart Grid*, 1(1), 2010.

267. A. K. Matias, N. Pincetic, G. Gross, A Successful Implementation with the Smart Grid: Demand Response Resources Contribution to the Panel: Reliability and Smart Grid: Public Good or Commodity, in *IEEE Power and Energy Society General Meeting*, July 2010.

268. P. Samadi, A. Mohsenian-Rad, R. Schober, V. W. S. Wong, J. Jatskevich, Optimal Real-Time Pricing Algorithm Based on Utility Maximization for Smart Grid, in *First IEEE International Conference on Smart Grid Communication*, October 2010, pp. 415–420.

269. J. Lee, D. K. Jung, Y. Kim, Y. W. Lee, Y.-M. Kim, Smart Grid Solutions, Services and Business Models Focused on Telco, in *IEEE/IFIP Network Operations and Management Symposium Workshops*, 2010, pp. 323–326.

270. D. S. Kerschen, Demand-Side View of Electricity Markets, *IEEE Transactions on Power Systems*, 18(2), 520–527, 2003.

271. C. A. Canizares, Conditions for Saddle Node Bifurcations in AC/DC Power Systems, *International Journal of Power and Energy Systems*, 17(1), 61–68, 1995.

272. C. A. Canizares, On Bifurcations, Voltage Collapse and Load Modeling, *IEEE Transactions on Power Systems*, 10(1), 512–522, 1995.

273. M. A. Pai, P. W. Sauer, B. C. Lesieutre, Static and Dynamic Nonlinear Loads and Structural Stability in Power Systems, *IEEE Proceedings, Special Issue on Nonlinear Phenomena in Power Systems*, 83, 1562–1572, 1995.

274. C. A. Canizares, S. Hranilovic, Trans-Critical and HOPF Bifurcations in AC/DC Systems, in *Proceedings of Bulk Power System Voltage Phenomena III – Voltage Stability and Security*, Fairfax, VA, August 1994, pp. 105–114.

275. P. A. Lof, T. Smed, G. Anderson, D. J. Hill, Fast Calculation of a Voltage Stability Index, *IEEE Transactions on Power Systems*, 7(1), 56–64, 1992.

276. B. Gao, G. K. Morison, P. Kundur, Voltage Stability Evaluation Using Modal Analysis, *IEEE Transactions on Power Systems*, 7(4), 1529–1542, 1992.

277. G. K. Morison, B. Gao, P. Kundur, Voltage Stability Analysis Using Static and Dynamic Approaches, *IEEE Transactions on Power Systems*, PWRS-8(3), 1159–1171, 1993.

278. N. D. Hatziargyrion, T. Van Cutsen, *Indices Predicting Voltage Collapse including Dynamic Phenomena*, Technical Report TF 38-02-11, CIGRE Taskforce, July 1994.

279. I. Dobson, Observation on the Geometry of Saddle Node Bifurcations and Voltage Collapse in Electric Power Systems, *IEEE Transactions on Circuits and Systems I*, 39(3), 240–243, 1992.

280. S. Greene, I. Dobson, F. L. Alvarado, Sensitivity of Loading Margin to Voltage Collapse with Respect to Arbitrary Parameters, *IEEE Transactions on Power Systems*, 12(1), 262–272, 1997.

281. P. Kessel, H. Glavitsch, Estimating the Voltage and Stability of a Power System, *IEEE Transactions on Power Delivery*, PWRD-1(3), 346–354, 1986.

282. S. Dey, C. K. Chanda, A. Chakrabarti, Development of Global Voltage Security Indicator (VSI) and Role of SVC on It in Longitudinal Power Supply (LPS) System, *Electric Power Systems Research*, 68, 1–9, 2004.

283. D. Thukaram, K. Parthasarathy, H. P. Khincha, N. Udupa, A. Bansilal, Voltage Stability Improvement: Case Studies of Indian Power Networks, *Electric Power Systems Research*, 44, 35–44, 1998.

284. C. K. Chanda, S. Dey, A. Chakrabarti, A. K. Mukhopadhyay, Determination of Bus Security Governed by Sensitivity Indicator in a Reactive Power Constraint Longitudinal Power Supply (LPS) System, *Indian Journal of Engineering and Materials Sciences*, 9, 260–264, 2002.

285. T. Nagao, K. Tanaka, K. Takenaka, Development of Static and Simulation Programs for Voltage Stability Study of Bulk Power System, *IEEE Transactions on Power Systems*, 12(1), 273–281, 1997.

286. N. K. Bose, P. Liang, *Neural Networks Fundamentals with Graphs, Algorithms and Applications*, Tata McGraw-Hill, India, 1998.

287. N. P. Padhy, *Artificial Intelligence and Intelligent Systems*, Oxford University Press, Oxford, 2005.

288. N. G. Hingorani, L. Gyugyi, *Understanding FACTS: Concept and Technology of Flexible AC Transmission System*, Institute of Electrical and Electronic Engineers, New York, 2000.

289. IEEE/CIGRE, *FACTS Overview: Special Issue*, 95TP108, IEEE Service Center, Piscataway NJ, 1995.

290. E. Acha, C. R. Fuerte-Esquivel, H. A. Perez, C. A. Camacho, *FACTS: Modeling and Simulation in Power Networks*, John Wiley & Sons, New York, 2004.

291. H. A. Perez, E. Acha, C. R. Fuerte-Esquivel, Advanced SVC Models for Newton Raphson Load Flow and Newton Optimal Power Flow Studies, *IEEE Transactions on Power Systems*, 15(1), 129–136, 2000.

292. C. R. Fuerte-Esquivel, H. A. Perez, E. Acha, A Thyristor Controlled Series Compensator Model for the Power Flow Solution of Practical Power Networks, *IEEE Transactions on Power Systems*, 15(1), 58–64, 2000.

293. D. P. Kothari, I. J. Nagrath, *Power System Engineering*, 2nd ed., Tata McGraw-Hill, India, 2008.

294. M. W. Mustafa, A. F. A. Kadir, A Modified Approach for Load Flow Analysis of Integrated AC-DC Power Systems, *Proceedings of TENCON*, 2, 108–113, 2000.

295. A. Konar, *Computational Intelligence: Principles, Techniques and Applications*, Springer, Berlin, 2005.

296. J. H. Holland, *Adaptation in Natural and Artificial Systems*, University of Michigan Press, Ann Arbor, 1975.

297. Y. Shi, R. C. Eberhart, Parameter Selection in Particle Swarm Optimization, in *Evolutionary Programming VII*, Lecture Notes in Computer Science, Springer, Berlin, 1998, pp. 591–600.

298. J. D. Bagley, The Behavior of Adaptive Systems Which Employ Genetic and Correlative Algorithms, PhD thesis, University of Michigan, Ann Arbor, 1967.

299. J. H. Holland, K. J. Holyoak, R. E. Nisbett, P. R. Thagard, *Induction: Processes of Inference, Learning, and Discovery*, Computational Models of Cognition and Perception, MIT Press, Cambridge, MA, 1986.

300. J. Kennedy, R. Eberhart, A New Optimizer Using Particle Swarm Theory, in *Proceedings of 6th International Symposium on Micro Machine and Human Science*, Nagoya, October 1995, pp. 39–43.

301. S. Das, A. Abraham, A. Konar, Particle Swarm Optimization and Differential Evolution Algorithms: Technical Analysis, Applications and Hybridization Perspectives, *Studies in Computational Intelligence (SCI)*, 116, 1–38, 2008.

302. D. G. Hart, Using AMI to Realize the Smart Grid, in *Proceedings of IEEE Power and Energy Society General Meeting – Conversion and Delivery of Electrical Energy in the 21st Century*, 2008, pp. 1–2.

303. S. M. Amin, B. F. Wollenberg, Toward a Smart Grid: Power Delivery for the 21st Century, *IEEE Power and Energy Magazine*, 3(5), 34–41, 2005.
304. D. Divan, H. Johal, A Smarter Grid for Improving System Reliability and Asset Utilization, in *Power Electronics and Motion Control Conference*, August 2006.
305. Federal Energy Regulatory Commission, Smart Grid Policy, Docket PL09-4-000, July 16, 2009.
306. S. Sen, A. Chakrabarti, S. Sengupta, ANN Based Improvement of Voltage Status of Weak Bus in a Multi-Bus Power Network Using SVC, *International Journal of Power Engineering*, 1(2), 153–164, 2009.
307. S. Sen, S. Sengupta, S. Chanda, A. Chakrabarti, Voltage Profile and Loss Assessment of a Power Network under Stressed and Contingent Condition Using TCSC and HVDC Link in a Deregulated Environment, *International Journal of Engineering Research and Technology*, 3(3), 467–480, 2010.
308. S. Sen, P. Roy, S. Sengupta, A. Chakrabarti, Generator Contribution Based Congestion Management Using Multi-Objective Genetic Algorithm, *Telkomnika: Indonesia Journal of Electrical Engineering*, 9(1), 1–8, 2011.
309. S. Sen, S. Sengupta, A. Chakrabarti, Alleviation of Line Congestion Using Multi-Objective Particle Swarm Optimization, *International Journal of Electronic and Electrical Engineering*, 4(1), 123–134, 2011.
310. S. Sen, S. Chanda, S. Sengupta, A. Chakrabarti, Differential Evolution Based Multi-Objective Optimization of a Deregulated Power Network under Contingent State, *International Journal on Electrical Engineering and Informatics*, 3(1), 118–131, 2011.
311. S. Sen, P. Roy, S. Sengupta, A. Chakrabarti, Genetic Algorithm Based Cost-Constrained Re-Dispatching Schedule in Deregulated Power Network, *International Journal of Electrical Engineering*, 4(3), 363–374, 2011.
312. S. Sen, S. Chanda, S. Sengupta, A. De, A. Chakrabarti, Swarm Intelligence Based Utility Optimization in Smart Grid Arena Employing Demand Response, *International Journal of Advances in Science and Technology*, 3(2), 25–34, 2011.
313. S. Sen, S. Chanda, S. Sengupta, A. Chakrabarti, Swarm Intelligence Based Congestion Constrained Load Curtailment Strategy, *ELECTRIKA – Journal of Electrical Engineering*, 14(1), 6–14, 2012.
314. S. Sen, P. Roy, S. Sengupta, A. Chakrabarti, AI Based Break-Even Spot Pricing and Optimal Participation of Generators in Deregulated Power Market, *Journal of Electrical System*, 8(2), 226–235, 2012.

Appendix A: Description of Test Systems

A.1 IEEE 14 Bus System

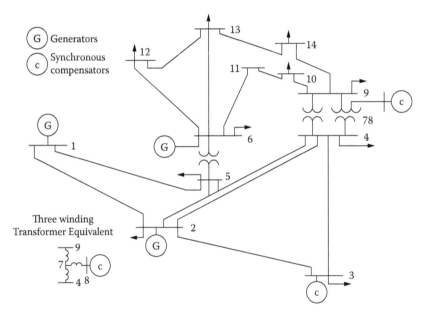

FIGURE A1.1
SLD of IEEE 14 bus system.

A.2 IEEE 30 Bus System

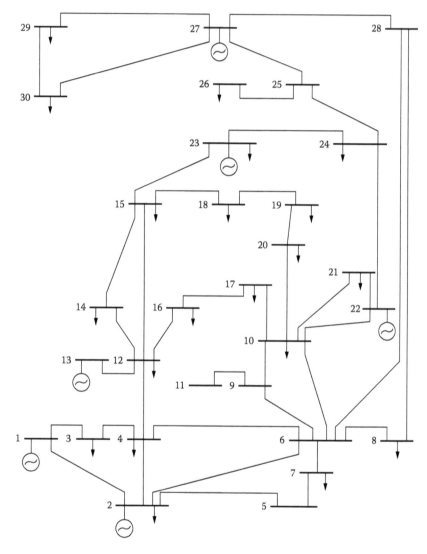

FIGURE A1.2
SLD of IEEE 30 bus system.

A.3 Eastern Grid of India

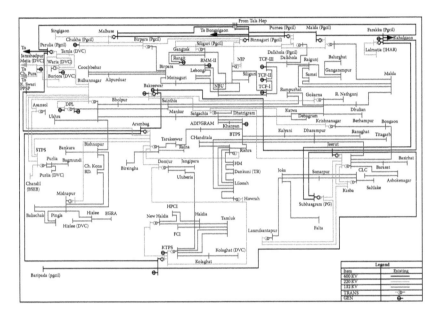

FIGURE A1.3
SLD of eastern grid of India.

A.4 Generator Cost Coefficients of the IEEE 30 Bus System

TABLE A1.A

Generator Cost Coefficients of IEEE 30 Bus System

| Bus No. | Active Power Output Limit of Generator (MW) | | Cost Coefficient | | |
	Min	Max	α	β	γ
1	50	200	0.00375	2.00	0
2	20	80	0.01750	1.75	0
5	15	50	0.06250	1.00	0
8	10	35	0.00834	3.25	0
11	10	30	0.02500	3.00	0
13	12	40	0.02500	3.00	0

Appendix B: Development of System Performance Indices

B.1 Development of Value of Lost Load (VOLL) (5.47)

VOLL can be defined as value of load loss due to unexpected contingencies in the power network. It is the product of p.u. generation cost after the required amount of load curtailment and the amount of load loss in the network.

$$\text{p.u. generation cost} = \frac{\displaystyle\sum_{i=1}^{NG} C_i^j}{\displaystyle\sum_{i=1}^{NG} P_{Gi}} \$/\text{MWh}$$

$$\text{Amount of load loss} = \left(P_i^0 - LCP^j\right)\text{MW}$$

Thus, VOLL can be developed as follows:

$$VOLL = \frac{\displaystyle\sum_{i=1}^{NG} C_i^j}{\displaystyle\sum_{i=1}^{NG} P_{Gi}}\left(P_i^0 - LCP^j\right)\$/\text{h}$$

B.2 Development of Value of Congestion Cost (VOCC) (5.48)

Total power generation cost in the developed method (multi-objective congestion-constrained cost and load curtailment optimization) is the summation of generation cost in the developed method and VOLL. Thus,

$$\text{Total generation cost/MWh (developed method)} = \frac{\displaystyle\sum_{i=1}^{NG} C_i^j + VOLL}{\left(P_i^0 - LCP^j\right) + \left(\displaystyle\sum_{i=1}^{NG} P_{Gi}^0 - P_i^0\right)}$$

Total generation cost/MWh (conventional cost optimization method) $= \dfrac{\sum\limits_{i=1}^{NG} C_i^j}{P_i^0}$

Thus, value of congestion cost can be developed as follows:

$$VOCC = \left[\frac{\sum\limits_{i=1}^{NG} C_i^j + VOLL}{\left(P_i^0 - LCP^j\right) + \left(\sum\limits_{i=1}^{NG} P_{Gi}^0 - P_i^0\right)} - \frac{\sum\limits_{i=1}^{NG} C_i^j}{P_i^0} \right] \times \left[P_{ij}^j - P_{ij}^0 \right] \$/h$$

B.3 Development of Value of Excess Loss (VOEL) (5.49)

Additional loss with regard to developed method (multi-objective congestion-constrained cost and load curtailment optimization)

$$= \left\{ \left(\sum\limits_{i=1}^{NG} P_{Gi}^0 - P_i^0 \right) - \left(\sum\limits_{i=1}^{NG} P_{Gi}^j - \left(P_i^0 - LCP^j \right) \right) \right\} MW$$

Thus, value of excess loss can be developed as follows:

$$VOEL = \frac{\sum\limits_{i=1}^{NG} C_i^j}{\sum\limits_{i=1}^{NG} P_{Gi}} \left\{ \left(\sum\limits_{i=1}^{NG} P_{Gi}^0 - P_i^0 \right) - \left(\sum\limits_{i=1}^{NG} P_{Gi}^j - \left(P_i^0 - LCP^j \right) \right) \right\} \$/h$$

Index